Collector's Guide to Vintage COIN MACHINES

Richard M. Bueschel

With Price Guide

77 Lower Valley Road, Atglen, PA 19310

Dedicated to...

My wife, Helen, who put up with coin-op "junque" for 30 years, and finally learned to like it and the people involved.

FRONT COVER PHOTOS

Bottom Left: Advertising Scale MR. PEANUT, 1949. Hamilton Scale Company of Toledo, Ohio, had a great idea. Make some of their PW "LowBoy" scales into advertising figures. They formed Advertising Scale Company division, and sold some major consumer products on the idea. Some were in the shape of pop bottles for ROYAL CROWN COLA, GRAPETTE and others. Most famous is MR. PEANUT made for Planters Peanuts. Very valuable. But watch it. This scale has been reproduced from "LowBoy" models. $15,100.

Top Left: Prophetron ZOLTAN, 1969. Often, in arcade machines, everything old is new again. The Lady-In-The-Box machines that started with the Roovers MADAME ZITA spawned a new generation of boxed figure fortune tellers every decade, well into the video era. Many of these latter day machines look a lot older than they actually are. The Prophetron, Inc. ZOLTAN for one. This fortune teller is a favorite for TV and movie sets that incur the past, often miscast in time. $925.

Top Center: Wurlitzer 1015, 1946. The most classic jukebox of all time, and the one best known throughout the world. The power of numbers, and advertising, made that possible. Nothing exceptional, the 1015 was the logical extension of Paul Fuller's design work. But the machine came out after four years of war with no new machines and met an enormous pent up demand. Well over 50,000 were produced, with sales driven by a national ad campaign. Fame came as "The Bubbler." $7,525.

Top right: Mills Novelty 1925 FRONT O.K., 1925. The classic Mills middle '20s front vender Bell machine, nicknamed "F.O.K." as a contraction of its model name. The show window glass is flat, and the anti-slugging viewing window at the right below the coin chute has been enlarged. As for dating this machine its serial is 123,758 in 5¢ play, standard coinage for an F.O.K., although 25¢ models were also made. The roll mint vending feature was used to get around the law. $1,225.

Bottom Right: Drobisch STAR ADVERTISER, 1897. Allows the advertiser, in this case a cigar called Star Advertiser made by the W.L. Kline Company in St. Louis, Missouri, to put their identification on the front of the machine. A no blank machine, the front panel promises "You always get one, and sometimes more." A pointer spins to 1, 2 or 5. A greater variety of Drobisch cigar wheels were produced in a few years than by any other maker. However, they remain fairly rare. $1,425.

Library of Congress Cataloging-in-Publication Data

Bueschel, Richard M.
    Collector's guide to vintage coin machines: with price guide / Richard M. Bueschel.
        p. cm.
    Includes bibliographical references and index.
    ISBN 0-7643-0579-4
    1. Slot machines--Collectors and collecting. I. Title.
    TJ1557.B84   1995
    668.7'52--dc20   94-48588
                                    CIP

Revised price guide: 1998
Copyright © 1995 by Richard M. Bueschel

All rights reserved. No part of this work may be reproduced or used in any forms or by any means graphic, electronic or mechanical, including photocopying or information storage and retrieval systems without written permission from the copyright holder.

ISBN: 0-7643-0579-4
Printed in Hong Kong
1 2 3 4

Published by Schiffer Publishing Ltd.
4880 Lower Valley Road
Atglen, PA 19310
Phone: (610) 593-1777; Fax: (610) 593-2002
E-mail: Schifferbk@aol.com
Please write for a free catalog.
This book may be purchased from the publisher.
Please include $3.95 for shipping.

In Europe, Schiffer books are distributed by
Bushwood Books
6 Marksbury Avenue
Kew Gardens
Surrey TW9 4JF England
Phone: 44 (0)181 392-8585; Fax: 44 (0)181 392-9876
E-mail: Bushwd@aol.com

Please try your bookstore first.
We are interested in hearing from authors
with book ideas on related subjects.

Acknowledgments

One person cannot speak for an entire hobby; the hobby must speak for itself. And the hobbyists are its best voice. I had the benign, aggressive and always enthusiastic support of hundreds, and probably thousands, of people who have found happiness and enlightenment in the finding, researching, restoring and the sheer enjoyment of collecting coin machines of every ilk. Some were more supportive and contributing than others, and I would like to specifically thank them here.

My first appreciation goes to the coin machine industry itself. I am afraid that I made a pest of myself for years among those whose job it was to invent, create, produce and market the temporal machines that collectors now treasure. They had a job to do to keep the wheels of their industry turning, yet they took the time to answer my questions and preserve their thoughts, and often documents, which now add substance to the history of these wonderful machines. Others shared their active years with me in their well earned retirement, at a time when future collectibility was the farthest thing from their minds. Without their willingness to cooperate, much of the story of the development of coin machines would have been lost.

After these individuals come the collectors. Few situations are as competitive; we were all looking for the same things. Yet they willingly shared their thoughts, exchanged data, and offered help. You will see a synthesis of their holdings, understandings, wisdom and advice in the pages following. An even more physical contribution has been made in the wealth of pictorial data presented on these pages. An agglomeration of vintage coin machines has little in common with a button or marble collection, where whole presentations can be put in trays and photographed with some ease once a light box has been set up. Not so for coin-ops. These machines tend to be big (and heavy!), with some a lot bigger than others. So every photograph requires a setup, not to mention the machine. Economics being what it is, there is no way that an itinerant hobbyist photographer could have shot all the pictures needed for this book without taking years and spending a fortune. The photography had to come from the machine owners, and from picture taking *in situ* at shows and shops. I asked, the pictures came, and I shot.

For graphics appearing throughout the book I particularly thank Rick Akers, Oklahoma City OK; John Apostolos, Milwaukee WI; Charlie Ball, North Little Rock AR; Burton A. Burton, Malibu CA; Larry De Baugh, Baltimore MD; Charles Deibel, Jeanette PA; Lee Dunbar, Dunbar's Gallery, Milford MA; Mme. Emmanuelle, Combs, France; Marshall Fey, Reno NV; Jeff Frahm, St. Louis MO; Jack Freund, Springfield WI; Bernie Gold, Great Neck NY; Mike Gorski, Westlake OH; Tom Gustwiller, Ottawa OH; Bill and Rosanna Harris, Denver CO; Stan Harris, Philadelphia PA; Bill Johnson, Rosamonde CA; Jack Kelly, St. Joseph MI; Rodger Knutson, Seattle WA; Bob Klepner, Victoria, Australia; Bernie Lasiewicz, Chicago IL; Larry Lubliner, Highland Park IL; Paul Olson, Sartell MN; Allan Pall, River Forest IL; Glenn Peterson, Houston TX; Don Reedy, Frederick MD; Ken Rubin, Brooklyn Heights NY; Alan Sax, Long Grove IL; Bill Triola, Lansing MI; Tom Venizelos, Pacifica CA; Ira Warren, Northridge CA; Jim Waters, Madison WI and Bill Whelan, Daly City CA.

Appreciation for specific graphics in the various chapters goes to (Slot Machines) Art Anderson, Flint MI; Charles F. Bass, Dublin NH; Greg Beresik, Pittsburgh PA; Bill Bernard, Salinas, CA; Ralph Bounty, Stamford CT; John F. Buegel, Grand Forks ND; Howard Culp, Worland WY; Jez Darvill, Surrey, England; Tom Erickson, St. Paul MN; Eugene Feack, La Quinta CA; Ray Francis, Pleasanton CA; J. P. Goosman, Ontario, Canada; Walter E. Gosden, Floral Park NY; Bud Gott, Eagan MN; Steve Grimando, Kew Gardens NY; Butch Hampton, Woodbridge VA; Harrah's Museum, Reno NV; Ron Hinze, St. Paul MN; John Herook, Lynfield MA; Sam Henka, Marina Del Ray CA; Michael Jacobson, Brooklyn NY; Stan Jankowski, Midland MI; Don Johnson, Wilmington DE; Richard Kahn, Woodland Hills, CA; Steven Kaplan, Penbroke Pines FL; Jim Krughoff, Downers Grove IL; Rick Lee, Lincoln MA; Bernie Madrid, Albuquerque NM; Dan Maggitti, Clifton Heights PA; Ron Maslowski; Tim McGovern, Scottsdale AZ; Jack Moneta, Shelburne VT; Tim Morsher, Kelly Island OH; John Nagel; Michael Olivas, Boynton Beach FL; Steve Patience, Pleasanton CA; Bill Price, Pittsburgh PA; Ron Pusateri, CA; Jasper Sanfilippo, Barrington Hills IL; Jim Saulati, Granada Hills CA; George Schultze, East Hampton NY; Peter Segara Photographer, Los Angeles CA; Ted Senness, Apple Valley CA; Ed Smith, Pecatonica IL; John Soulsby, Madison OH: Bob Stone, Reading MA; Gary Sturtridge, Tonganoxie KS and Jim T., Lynn MA; (Juke Boxes) Rob Berk, Warren OH; Patrick J. Bilotta, Newark NY; Mike Hammack, Chesterfield VA; Charles de Hares, Los Angeles CA; John Fisher, Carlsbad CA; Kenneth M. Glover, Houston TX; John T. Johnston, Brooklyn NY; Richard Litsinger, Potomac MD; Don Marek, Grand Rapids MI; Caesar Milch, Tujunga CA; Stan Muraski, Rockford IL; John Papa, Mayfield NY; Jimmie L. Rice, Wichita KS; Bob Sandage, Louisville KY; Steve Siran, South Elgin IL; Hank Sontag, Buffalo NY; John Wilcox, Oroville CA; (Pinball Games) Tom Blum, Kalamazoo MI; Tim Brady, Wichita KS; Rich and Lois Carroll; Richard Conger, Sebastopol CA; Bill Cowles, Norwalk CA; John Detrisson, Dundee NY; Phil Emmert, Colorado Springs CO; Greg and Judy Falletich, Westminister CA; Les Hahn, West Bend WI; Pat Hamelet, Chicago IL; Sam Harvey, Pomona CA; David Haynes, Santa Rosa CA; Sharon Herren and David Humphveville, Houston TX; Tim Huntley, Guernsey, Channel Islands, U. K.; Russ Jensen, Camarillo CA; Jim Marescalco, Kenosha WI; Michael Shahnamian, Babylon NY; Jeff Mischke, Cheyenne WY; David L. Wright, Stow OH; Steve Young, Lagrangeville NY; (Arcade Machines) Carl Ancman, San Antonio TX; Donald O. Beckwith, Honolulu HI; Jack Biehl, Mt. Clements MI; Barry B. Boswell, Aurora CO; Nic Costa, London, England; William P. Daugharty, Saginaw MI; Gary Dressel, Shorewood MN; Ken Durham, Washington DC; Robert Ferguson, Mt. Prospect, IL; Steve Gronowski, Barrington Hills, IL; Rick Hoffman, York PA; Bill Jenkins, Utica NY; Tim LaGanke, Novelty OH; Bob Legan, Euclid OH; Pete and Pam Movsesian, Fountain Valley CA; Marty Roenigk, East Hampton CT; Herb Silvers, Los Angeles CA; Roger Simale, Chicago IL; Mark Weyna, Des Plaines IL; Carlton Wilkins, Bloomfield CT; (Trade Stimulators and Counter Games) Barton Battaile, Lexington KY; Bill Howard, Akron OH; Barbara and Bill Huber, Greenville OH; Joseph R. Jakubuwski, Ponta Gorda FL; Walter K. Mintel Jr., Madison NY; Kent Patterson, Edwardsville IL; Rich and Sharon Penn, Waterloo IA; Dale Rollins, San Antonio TX; Joe Welch, San Bruno CA; Frank Zygmunt, Westmont IL; (Vending Machines) Rod Brandt, Normal IL; Lisa Bremerkamp, Ft. Wayne IN; Eugene Brown, Burbank CA; John Capezzuto, Pittsburgh

PA; Louis Chomyk, Auburn NY; Bill Enes, Lenexa KS; Paul S. Firman, Woodbury NJ; Robert Fox, West Point OH; Stephen W. Karthes, Bothwell WA; Rev. Edward and Sandra McNulty, Dayton OH; Denny Roeck, Dayton OH; R. Graham Shaw, Huntingdon PA; Dennis Valdez, Hercules CA; (Coin-Operated Scales) Robert Batto, Rochester NY; Bill and Jan Berning, Sycamore IL; Rick Crandall, Ypsilanti MI; Fairbanks Museum and Planetarium, St. Johnsbury, VT; Jon Gresham, Penny Arcadia, York, England; Al Guerra, Ridgewood NY; coin machine historian and museum curator Birgit Friederike Haberbosch, Lubbecke, Germany; Roger Harper, New Albany, IN; Ron Heatley, Sevierville TN; Barbara Larks, Skokie IL; Gerald "Red" Meade, Los Angeles CA; Gary Moise, Orange MA; Charles Rizzo, Kenosha WI; Larry Sweeney, Louisville KY; Dale Wilson, Flossmoor IL and Pete Wojtala, Trenton MI.

Pace ALL STAR COMET GOLD AWARD SKILL, 1936. Adding a Gold Award to the ALL STAR COMET changes its front and color. The GA models were painted orange with blue paint and chrome trim, earning the nickname "Orange Front." This is an export model in 1-franc play for France, adding a European and British option, skill stops to abruptly stop the reels when the buttons above are pushed. Serial is S-46,958-Gs, the "S" standing for ALL STAR and "GS" meaning Gold Award Skill. $1,525.

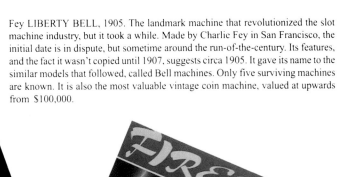

Fey LIBERTY BELL, 1905. The landmark machine that revolutionized the slot machine industry, but it took a while. Made by Charlie Fey in San Francisco, the initial date is in dispute, but sometime around the run-of-the-century. Its features, and the fact it wasn't copied until 1907, suggests circa 1905. It gave its name to the similar models that followed, called Bell machines. Only five surviving machines are known. It is also the most valuable vintage coin machine, valued at upwards from $100,000.

Bally FIREBALL, August 1971. Quite possibly the best known pinball game ever made. Bally was riding high as the leader producer in the industry when FIREBALL was created. The lower center playfield has a spinning disc that sends the balls in any and all directions. When it was featured in a story in *Playboy* it became world renowned overnight. High value for a pinball. This flyer is signed by game designer David G. Christensen. It is also valuable. $160.

Contents

Introduction....6
The Universe of Coin Machines....7
Slot Machines....9
Jukeboxes....69
Pinball Games....90
Arcade Machines....110
Trade Stimulators and Counter Games....136
Vending Machines....165
Coin-Operated Scales....183
Building a Collection....197
Resources....206
Index....215

Pace 1936 PACES RACES, 1936. Continual improvements, and new cabinets, kept PACES RACES in the forefront of slot machine activity for ten years, and even longer when it was re-made by another manufacturer, H. C. Evans, also in Chicago. The 1936 "Brown Cabinet" and subsequent models added features and jackpots. These machines continually increase in collectible value, with no end in sight. A spinning disc at the far end of the race track changes the odds as the race runs. $8,520.

Introduction

In the beginning they were called Automatic machines. That was in the early 1880s, when the idea of a machine that was actuated by a coin leapt across the Atlantic Ocean from the British Isles and took a firm foothold in the United States of America. The leap was human, because the opportunity for a big financial score in England was limited by the enormous competition anybody with talent, ability, and skill but without financial backing faced. Not true in the U.S. Here was a raw nation just beginning to feel its oats (and wheat, and barley, and pig iron) that was embarking on its biggest building program ever. Take a look at the old store fronts on Main Street in any town, or the cornerstone in any old town hall, county court building, or railroad station and note how many were built between 1880 and 1890. Lots. The opportunities that expansion presented led to an instant British brain drain. So Thomas (later "Sir") Crapper came over here to sell flush toilets to city halls (and left his name on the product, and subsequently the room that held it); Montague Redgrave came to Cincinnati (an early bird; he came in the 1870s) to invent the spring plunger marble shooter that led to a significant American industry in bagatelle games that lasted over 60 years, eventually evolving into the pinball game; Percival Everitt leaned on family connections and came to St. Johnsbury, Vermont, in 1885 to sell the powerful and well-financed E. & T. Fairbanks Company (one of America's leading exporting companies before the 1900s) on the idea of an automatic scale (Fairbanks records don't mention it, but one of the original Everitt scales of the 1880s was recently found in the basement of the St. Johnsbury Museum, and it's a coin-op), and opened up a company in New York City the next year to promote penny postcard venders at street railway stations; a Londoner named William S. Oliver attempted to sell his 1887 talking (phonograph) shock machine, only to have it copied by The Automatic Battery Company in Brooklyn in 1888; and a man named Anthony Harris from Middlesborough came to New York in 1888 to interest American investors in a chance machine, selling the idea but seeing at least a dozen imitators by the end of the decade. So by 1890, the American forms of pinball games, trade stimulators, arcade machines, vending machines, coin-operated scales, automatic phonographs (later to be called jukeboxes) and finally the automatic-payout slot machine were well on their way to domestic and, soon to come, international acceptance as American ambassadors to the world. For over 80 years the American coin-ops held reign, and it has only been in the past decade or so, since the advent of Japanese-originated arcade games, that this commanding position has been threatened. That's a period of four generations of production and subsequent collectibility, for many of these vintage machines are still being found and have formed the basis for major collections. By the 1890s, and increasingly so, the word "Automatic" fell out of favor for the coin-ops for the simple reason that it did not explain the class of machine. In fact, *most* new machine inventions between 1880 and 1920 (and certainly thereafter) were automatic in one way or another. So the industry came up with its own name. In the 1890s the descriptive appellation "Coin Controlled Machines (CCM)" was applied, and many dated city directories list the makers under such a heading. By 1900, and the beginning of The New Century, the simplified description "Coin Machines" took hold.

So much for the origins of coin machines. The purpose of this book is to bring some of the charm and excitement of coin machine history, identification, collecting and actual machine finding to a broader scope of enthusiasts to encourage them to become aware of the select group of 5,000 or more avid collectors of these machines in the Americas, and perhaps 6,000 worldwide. It is an enthusiasm that didn't exist in the 1950s, but has grown by leaps and bounds since the '60s. Share in its joys, and join us.

Richard M. Bueschel
414 N. Prospect Manor Avenue
Mt. Prospect, IL 60056-2046
Tel. 1-708-253-0791
August 1, 1994

The Universe of Coin Machines

Coin machines have been made throughout the industrialized world for a little over a century, with development and production continuing domestically and offshore while experiencing popularity ebbs and flows. The machines are often a direct reflection of popular culture, and frequently exhibit graphics and themes that are taken from other media, such as entertainment, as well as the news events of the day. As such they are a continuing reflection of their times. While historically the machines started in England, that 1880s leap across the Atlantic to American shores gave them a new home. Since the 1890s, they have largely been an American industry. Until slot machine manufacturing was dispersed by The Johnson Act, outlawing gambling machines in January 1951, most of the machines of all classes came out of Chicago. The Windy City became known as the "Coin Machine Capital of the World." Chicago is still the leading producer of pinball games and was for payout slot machines, until the formation of International Game Technology in Reno, Nevada and the recent Bally Gaming move west to join IGT.

If I didn't make the point that this is a broad subject, I should. Over the years more than 8,500 different pinball games have been made; 3,500 plus different slot machines; unnamed thousands of arcade machines, with guesses at about 9,000; at least a thousand different scales; an equally large number of jukeboxes and trade stimulators; and an uncharted number of vending machines estimated at over 10,000 different models. So we are looking at a universe filled with over 30,000 different coin machines made in the past century. And that's only the American production. Offshore assembly—British, French, German, Japanese, Australian, Italian and Spanish, to cover the main producers—will probably add another 10,000, so we get into really substantial numbers.

Of the big name full line producers located in Chicago, the unrivaled leader for the first half of the century was the Mills Novelty Company (1897-1953). Mills was trailed by many other major producers including: Bally Corporation (1931-to date), now Bally Gaming out of Nevada; O. D. Jennings & Company (1907-to date) starting as Industry Novelty Company, becoming Jennings in 1921, and still surviving in name as half of the Mills Jennings Company in Nevada; Watling Manufacturing Company (1902-1963) and the Pace Manufacturing Company (1926-1953) to name a few. These were followed by a number of other Chicago based vending machine and pinball producers such as Advance, Chicago Coin, Exhibit Supply, Gottlieb, Williams, A.B.T. and numerous others.

Few of the significant coin machine manufacturers of the past were single line producers. The big manufacturers all produced just about everything, from scales to slots to pinballs and arcade machines, not to mention vending and trade stimulators and jukeboxes. A look at the product mix reveals the complexity of the company offerings, with only some of the vending machine companies restricted to single product lines. Further variation in productive capability is found in the type of payout slot machines being made, with some firms only making the traditional mechanical 3-reel counter models while others made consoles, or larger floor standing machines that generally operated by electricity. Other makers restricted their slot machine capability to the upgrading and revamping of existing machines (usually starting with a Mills machine), such as Rock-Ola in Chicago. Some of the revampers advanced beyond that to create their own machine lines, a prognosis followed by Silver King in Indianapolis, Indiana; Superior Confection Company in Columbus, Ohio; and Buckley Manufacturing Company in Chicago. It was only with the introduction of the electromechanical consoles in the late 1940s (produced throughout the '50s and leading to the revolutionary Bally electromechanical slot machines of the '60s and '70s to the present electronic and digital era) that the industry began to truly specialize by manufacturer. Bally abandoned their vending machines and arcade kiddie rides to concentrate on slot machines and pinball games. Other Chicago makers such as Williams, Gottlieb (later Mylstar, to be bought out by Premier Technology in 1984), Exhibit Supply, Chicago Coin, and Genco restricted their offerings to pinball games in the 1950s and '60s. These firms either went out of business or added arcade machines in the '70s and '80s, with the survivors joined by new makers in the '80s and '90s. Jukebox producers such as Wurlitzer out of North Tonawanda, New York, and the cadre of Chicago producers including Seeburg, Rock-Ola, Mills and others also edged into other fields. Wurlitzer made a few arcade games, while Seeburg and Rock-Ola made a variety of models, including pinball games and arcade machines. The ebb and flow of coin machines continues, making machine history today as it did in the past, with tomorrow's collectibles still moving off of the shipping docks of the new and surviving coin machine makers of the present. Only one maker, Williams Electronics in Chicago, shows signs of emulating the full line producers of the past with various divisions making pinball, arcade, trade redemption, and payout slot machines—the classic product mix of bygone years. Few consumer entertainment and service industries are as enduring, and there is little question that major collectibles of the future are yet to come. Only we don't know which ones they will be. Only time will tell, just as time has earmarked the machines shown on the following pages.

Author Richard M. Bueschel of Mt. Prospect, Illinois. Dick is holding a restored American Automatic Machine Company AUTOMATIC DICE SHAKING MACHINE trade stimulator of 1892. In the foreground, to the far left, is one of the relatively rare initial production Bally 742A MONEY HONEY slot machines of 1963 with the formica "Whitesides" and fake double jackpot front, the latter quickly dropped to meet Nevada gaming requirements. To his immediate left is a cast iron Mills Novelty TARGET PRACTICE trade stimulator of 1918, behind which appears the 1904 franchise wall display from the Chicago office of the National Automatic Weighing Machine Company showing the eleven scales circa 1889-1904 available for placement. © Copyrighted (November 24, 1991), Chicago Tribune Company. All rights reserved. Used with permission. Photo by Hung Vu.

The 30 leading (in ranking order) vintage coin machine manufacturers

Company, City & Date			Slots	Trade/Counter	Arcade	Pinball	Jukebox	Vending	Scale
Mills Novelty	Chicago	1897-1953	X	X	X	X	X	X	X
Caille Bros.	Detroit	1897-1941	X	X	X	X	X	X	X
Bally	Chicago	1931-Present	X	X	X	X	X	X	X
O. D. Jennings	Chicago	1906-1979	X	X	X	X	X	X	X
Watling	Chicago	1902-1963	X	X	X	X	—	—	X
Pace	Chicago	1926-1953	X	X	X	X	—	X	X
Advance	Chicago	1904-1957	—	X	X	—	—	X	—
Chicago Coin	Chicago	1932-1978	X	X	X	X	X	—	X
Pacific	Chicago	1932-1941	X*	X	X	X	—	X	—
Wurlitzer	N. Tonawanda NY	1854-1979	—	—	X	X	X	X	—
Keeney	Chicago	1916-1974	X*	X	X	X	—	X	—
Exhibit Supply	Chicago	1917-1978	X*	X	X	X	X	—	—
Rock-Ola	Chicago	1927-Present	X**	X	X	X	X	X	X
Gottlieb	Chicago	1928-1984	X*	X	X	X	—	—	—
H. C. Evans	Chicago	1902-1957	X*	X	X	X	X	—	—
Western	Chicago	1934-1939	X*	X	X	X	—	—	—
Seeburg	Chicago	1899-Present	X*	—	X	X	X	X	—
Buckley	Chicago	1931-1962	X***	X	—	X	X	—	—
A.B.T.	Chicago	1927-1955	—	X	X	X	X	X	X
Williams	Chicago	1943-Present	X	X	X	X	X	—	—
Groetchen	Chicago	1931-1955	X	X	X	X	—	—	—
Daval	Chicago	1932-1953	—	X	X	X	—	X	—
United	Chicago	1945-1955	—	—	X	X	X	—	—
Superior	Columbus, OH	1930-1941	X***	X	X	X	—	—	—
Silver King	Indianapolis, IN	1919-1939	X**	X	—	—	—	X	—
Mutoscope	Long Island City, NY	1897-1952	—	X	X	X	X	—	(X)
Northwestern	Morris, IL	1918-1990	—	—	—	—	—	X	—
National Weigh	St. Paul, MN	1919-1941	—	—	X	X	—	—	X
AMI/Rowe	Grand Rapids, MI	1928-Present	—	—	—	—	X	X	—
Columbus	Columbus, OH	1905-1950	—	—	—	—	—	X	—

*Consoles only. ** Revamp only. *** Revamper, added own lines. (X) Distributor.

Slot Machines

When you come right down to it, the number one favorite vintage coin-op collectible is the automatic payout slot machine. That's right! Collectors favor the hard core gambling machine. This class has one of the longest lifetimes of any coin machines, starting around 1889 and remaining in continuous production until the present day; changing, evolving and developing all along the way. It's a lot of fun to have your own casino machine at home, and as long as you are working with your own money you can't lose. But the fact that a slot machine was created as a gambling device makes it subject to state law in the United States and to national law in other countries. It is something you should be aware of, for slot machine collecting—the automatic payout ones, where money comes out if you hit a "winner"—is governed by law. Time was, back in 1976, when it was illegal everywhere to collect slots, you could own one only in Nevada, and even there the laws were funny, so if a friend played a nickel, won, and took the money home you were illegal. Since then, with a lot of collector effort (and risk!) that situation has changed, making it more than likely that you live in an area where slot machine collectibility is legal, provided your machines are not used for public gambling.

And that is the content of the law. Permission to collect doesn't change the "organized crime" and gambling statutes of the 1950s and 1960s, but ownership of an antique slot is legal where the laws say "It can be a defense that" explaining the antique status and the fact that no one can come in and gamble on your machines. That's why true collectors have a glass of coins next to their treasures, so that you can play but keep the money there to be recycled. So it's not gambling. Otherwise, bam! You could be in deep trouble.

Currently, open antique slot machine collecting is allowed in 44 states. There are a few iffys. Hawaii, for instance. If your machine can play and pay, it's illegal. But if you jimmie it to make it inoperable as a gambling machine, it's okay. Nebraska and Indiana are tough to call. There are no laws that permit collectibility yet (they will come!), so there is a degree of jeopardy when displaying, selling, buying and allowing people in your home to play a slot. But for Alabama, Connecticut, South Carolina and Tennessee it's a solid no-no. Real trouble no-no. You can't do it! It's going to take collectors in those states with the courage to buck the system and try and get some collectible laws passed (exposing their interest in the process) before these unenlightened states will allow their citizens to partake in one of the most interesting and involving hobbies in the country (and increasingly, in the world). A continuing update of the laws has been undertaken by *Coin Drop International*, one of the vintage coin-op collectible magazines. They have a "Status of the States" story in every issue. You'll find this publication listed in the "Resources" chapter later in this book.

The tricky part is legally defining an antique or vintage slot machine. Antique cars are based on an age of 25 years or more, which means that a Ford MUSTANG is viable antique car and can get plates as such. Try as they might, the slot machine collectors made an effort to get "25-Year Laws" in most states, but some of the honorable legislators balked. Those states that had a fear of past gambling sometimes had a cutoff date of 1941, which means a machine has to be over 50 years old (and older as time goes by) to be classified as a collectible. Other states say any date is okay, as long as the machine is in a home and not available for gambling purposes, so you've got to watch your states and be up on these ever changing laws. (In fact, some of the "1941 Law" states have been upgraded to "25-Year Law" states and more are following as the collectors make lobbying efforts to get their collections recognized as the marvelous artifacts that they are). The growing tendency to casino, reservation, riverboat and other local gambling across the country is making many of these collectibility laws archaic documents of a bygone age, but until a law changes what's in effect is the law to be followed.

John Lighton SLOT MACHINE, 1892. The earliest known surviving automatic payout, called a 3-For-1 because the machine returned 3 nickels if the one dropped in the top went in the proper direction when it hit the pins in the oval shaped window in the upper center. Payout was made in the small cup at the far left. Engraved picture of the Lighton JOHN LIGHTON'S LIFTING MACHINE (see Arcade Machines chapter) is put on front of the machine to show its family heritage. $6,050.

All of this legal stuff has led to another major change in slot machine collecting, and that's the flood level surge of the high powered electromechanicals of the 1960s and 1970s, dumped by casinos in Nevada and Atlantic City as new solid state and stepper driven reel machines came on line. The Bally MONEY HONEY is a classic example. First made in 1964, the early models became "legal" (in "25-Year Law" or no restriction states only) in 1989, with the newer models moving into the collectibility area as time goes by. Many new collectors have never seen some of the old mechanical slots that collectors lived with and adored prior to 1989, and a recent survey by one coin-op publication showed that about 1/4 of the slot machine collectors regarded themselves as Bally "specialists."

To most vintage slot machine collectors, when you talk about slots you are most likely talking about the durable 3-reel mechanical Bell machines that all owe their origins to one Charles August Fey. Fey was a German (Bavarian) by birth who came to the United States from Der Vaterland. He traveled to California in a roundabout way, including stops in Hoboken, New Jersey and Fond du Lac, Wisconsin prior to settling in San Francisco in the early 1880s. Industrious, clever and highly inventive, he got involved in coin machines very early, making his first machine in 1894. Fey, and a man named Gustav F.W. Schultze, and another named Gustave Hochriem, really gave birth to the slot machine industry of today. True, there were coin machine makers in the East, mostly in Chicago, Illinois but also in Brooklyn, New York City, Elmira, Syracuse, and Rochester, New York, Columbus, Ohio, and Detroit, Michigan; but it was Schultze, Hochriem and Fey, each creating their own basic formats, that led to the automatic gambling payout machines that endured for years. It can probably be said that the automatic payout slot machine was a German invention of the late 19th century, only it was created in America, and specifically in California.

were made as countertop models, and soon in large floor cabinets with massive cast iron coin heads that permitted a wide variety of "bets" on selected colors utilizing a variety of payout methods. The most sophisticated was a finger probe and coin slide system used on a machine called the PUCK, made by the Illinois Machine Company in Chicago and The Caille-Schiemer Company in Detroit starting in 1898. It was a system in which a color wheel spins, with a corresponding punched wheel behind it. When the wheel stops, springs release probes behind it and if any poke through a hole in the disk a proper payout is made. Around 1904, Hochriem created a single reel machine with a similar disk and probe system he called the ELK that made token payouts from tubes. The ELK was produced in Chicago by a new company called Paupa & Hochriem. Both machine formats soon had national, even international, sales.

Jones Novelty SLOT MACHINE, 1893. The John Lighton 3-For-1 format was widely copied by other makers in new cabinets. This is the Jones Novelty Company version made in Rochester, New York in 1893 and 1894. Payout is in the small slotted trough in the base at the far left. The lower window showed the coins to be won, much like an early jackpot. $3,010.

John Lighton SLOT MACHINE NO.4, 1893. An improved saloon and cigar counter model with a larger payout cup, plus the addition of a kerosene lamp at right to create the first lighted slot machine. At left is a multiple use cigar lighter, alcohol cup, match cup and striker for smokers. The John Lighton Machine Company was located in Syracuse, New York. $7,525.

The Schultze format was a variable metered payout automatic color wheel slot machine developed in the early 1890s and copied in Chicago and, in an unlikely technology transfer, in Paris, France. The Chicago makers were the D. N. Schall Company, Paul E. Berger, the Cowper Manufacturing Company, and Charles A. Wagner. These spinning color wheel machines

Charlie Fey surpassed all of these developments with a 3-reel payout machine made around the turn-of-the-century, most likely around 1905. Fey married the 3-reel display of older countertop card machines and modified the finger probe and slide payout system of the PUCK, coupling the reels to three matching punched discs that made the machine a mechanical computer. This innovation led to the first modern automatic payout slot machine. No longer did you need to pick a color. The great variety of "shows" offered by the three reels made a series of variable payouts possible. The basic 3-reel format holds true to this day. Glad to be an American (while still drinking San Francisco steam beer with his displaced countrymen) Fey called his breakthrough machine the LIBERTY BELL. With it he revolutionized the modest slot machine business and provided a mass producible machine that would turn it into a major industry.

If you want a valuable slot machine, find a Fey LIBERTY BELL. It is unmistakable. It has little golden colored cast iron feet, a cracked Liberty Bell symbol on top, 3 reels with card playing symbols, an all metal black painted cabinet and the names "Chas.Fey & Co. S. F." in the top casting. You can't mistake it for any other machine. Yet hundreds of people do for the very simple reason that it is worth anywhere from $25,000 to $75,000 or $100,000 and even more depending on whether you are talking to a seller or a buyer. Four are known; maybe a 5th and even a 6th. Yet around a hundred were made as it's known serial numbers go almost that high.

Maley PYRAMID BANKER, 1893. The "Two-Door Bank" format was almost as widely copied as the 3-For-1 machines. Coins trickled down over pins and if they fell into one of the two small holes at the center bottom, one or the other of the "bank" doors flew open to offer a bunch of nickels as a payout. Charles T. Maley Novelty Company was in Cincinnati, Ohio. There were at least five other makers. $3,210.

Nonpariel Novelty NONPARIEL, 1893. Another 3-For-1 copycat machine, made by the Nonpariel Novelty Company in New Haven, Connecticut. The payout trough is located in the base at the right side. Note the differences in bases. There were at least four other makers of virtually identical machines. $3,510.

The problem is that just about everybody else who copied the format after that named their machines the LIBERTY BELL, or made similar machines called the OPERATORS BELL, put on the little feet, and kept the name long after wooden cabinets came along to make the machines lighter. The big maker was the Mills Novelty Company in Chicago, who started making their LIBERTY BELL in 1907. And then everyone else (Watling, Caille, Silver King, Industry Novelty, Burnham & Mills and some others) copied Mills. Or revamped—that is rebuilt—the Mills cast iron cabinet machines in their image until the coming of World War I when the wooden cabinet cast front machines finally entered the picture. That continued until the 1960s.

The odds of your finding or buying a Fey LIBERTY BELL are infinitesimally low. The odds are very high, however, that you'll be able to get some other fine vintage machine if you look for one. When you are considering buying or acquiring a slot machine you should be very cognizant of the "Five Re's." Drill them into your head: real, restored, revamped, replica and (un)realistic. I'll even add a 6th one: research. More mistakes are made in putting out good money (or trading) for coin-ops when the "Re's" aren't considered than in almost any other area of antiques. And there are reasons for that. It's very much a let-the-buyer-beware kind of market. Not so much because chiselers are out there (although there are some) but rather because both the seller and the buyer often don't know what they are dealing with. It's a situation faced in all mechanical antiques, from guns to cars to boats to vintage computers to anything that's supposed to work and do something and look good at the same time. While the rule of thumb suggests that you should look for a machine that matches the pictures in a price guide, the fact of the matter is "you get 'em like you find 'em." Somewhere between the two is what's actual, and how to make everything "satisfactual." And that's where the "Re's" come into play.

The first and foremost "Re" is reality. Is the machine you are looking at real? If it's old and grungy and has a wasp nest in it, that's about as real as you can get. But even then there are sometimes questions. Price guides seem to concentrate on the pristine examples of any machine, and a Mills SILENT "War Eagle" of 1931 looks one way, and no other. But consider this. The Mills MYSTERY "Blue Front" of 1933 had two jackpots whereas its later variant the BLUE FRONT "Castlefront" of 1937 only had a single, larger, jackpot. An example of the later version was also produced as the FUTURE PAY model, called the "Red Front" for its change in color. Yet when a two jackpot "Red Front" shows up, is it real? Well, that's the way the old operator ran the game, and if it operated that way it's valid, cataloged or not. So even if the game is incorrect, that doesn't mean it isn't real. Research (the 6th "Re") is important here.

But what if the machine you find is all shiny and has recently been replated? Is it real? Not to me. Maybe to some collectors, but as far as I see it, when you demean the integrity of the machine—the moment you make it something it wasn't, or add something that didn't run on the original—you have made it (un)realistic. Now, if the machine had been cleaned up, even repainted, but in true faith to the original, that's a restored machine. In restoration you can do a lot of things to make the game work, by adding new springs and even mechanical parts to replace broken components. But as long as you maintain the integrity of the machine in it's cosmetic, or appearance, mode, it's real, and restored.

Clawson THREE JACKPOT, 1893. An overnight saloon sensation, the big pot payouts of the THREE JACKPOT made this the most popular machine of its era. It established its maker Clement C. Clawson as the premier coin machine designer in the country. The Clawson Slot Machine Company was located in Newark, New Jersey. Push down at the top and get the coin into one of the three slots in the playfield, and the pot below dumps when you push the top plunger down a second time. $2,010.

Columbian Novelty SLOT MACHINE, 1893. Perhaps one of the most elegant 3-For-1 machines, the COLUMBIAN was named to coincide with the opening of the World's Columbian Exposition World's Fair in Chicago in May 1893. It was made by two brothers with a woodworking shop in East Saginaw, Michigan, named Adolph and Art Caille. In less than a decade they would become leaders in the soon to be booming slot machine industry. $3,520.

There are times that changes can be accepted in a real machine, but once again it's based on how the machine ran. When the mid-1920s slots slowed down in operation because the public got bored with the same old game and its small payouts, the jackpot was introduced. First, before dedicated jackpot machines were made, new jackpot fronts were added to the old machines. The old plain cast iron or aluminum machine front was taken off, and the new front replaced it, coupling the mechanism to trip the new higher payout when the proper symbols were hit. Other new fronts added skill features, baseball or football play features, and all sorts of newfangled ideas as they came along in the late 1920s through the 1930s. They are called revamps. The changes were real, it happened "then," and it changed the integrity of the machine. It's a very acceptable "Re." Unless it's all bright any shiny like any other overly restored (un)realistic restoration, then the prior "Re" rule applies.

Are you keeping up with this? Because if you are, here's more. Big bad "Re:" replica. Once the value of the antique slot machine went above a thousand dollars it became economically feasible to make replicas. While I almost hate to admit it, there is a place for replicas as well as overly restored machines. They are new or as-new, often work better than the vintage machines, and start right out looking great. If it's a game room machine you want, just to play around, a replica or an upgraded restoration may be the way to go. As long as you know that's what it is. But some of the replicas and restorations are so good these days they often pass for the real thing, and once they do that (or are sold that way) they're fakes. You can usually tell. The grubbiest components in a slot machine are the payout slides, because they get the most work. Take a look inside, and if the slides show wear and dirt and maybe they don't work too well, you've probably got the real thing. But if they are brand-spanking new silvery metal, or are made of plastic, watch it. You've probably got a replica on your hands.

Then there's the (un)realistic, or the existence of machines that seem to be the real thing but actually never existed in the past. These are fantasy items made up to look good (and they usually play very well, so they're fun) while they have absolutely no collectible value. None! The classic example is the GOLDEN NUGGET "half-top" slot machines that look like Mills Novelty revamps but in reality are usually replicas made up to look like something that somebody today thinks a slot machine should have looked like in the past. The Golden Nugget Casino actually ran a few that were similar back in the '50s and '60s, but they were Mills GOLDEN FALLS revamps and are all long gone. In their place are all sorts of new fantasy versions that even the Golden Nugget casino has complained about, asking the producers to refrain from making more. These fantasy machines will be around for years to come, and in the future may even be traded as valid antiques. But they aren't.

Amusement Machine 3 JACKPOT, 1893. The Clawson THREE JACKPOT became one of the most copied formats in slot machine history. Over a dozen makers produced similar machines between 1893 and 1910. The 3 JACKPOT model was made by the Amusement Machine Company in New York City, one of the first major chance machine producers. Half loop at top is anti-slam device to prevent players from hitting the plunger too hard. $1,810.

Finally, that last "Re:" research. Knowledge is certainly power, and a money saver and mistake preventer. With all the things we touched on leading up to research you can see the invisible hand of knowledge making major decisions, and that's the way it ought to be. Once you know your machines, and can spot them by their looks and details, you are pretty much on safe ground when it comes to buying or selling. But there are still traps out there for the unwary, such as offshore machines. The United States isn't the only country that made coin machines, and some of the British and European machines are truly unique and different. But face it, they aren't the collectibles most American collectors want. For one thing, the coinage isn't the same size as American change, and that should be a fast tip off to any offshore import. For another, most of them are "wall machines," meant to hang on a wall. This class of machine never—repeat, never!—made it to the American market. So if it hangs, it's offshore. There's another tip off, and that's the coinage. If the coin chute says PENNY, HALF-PENNY or 1d it's British, using the old penny; 1f is French for the franc; pf is German for pfennig. In the main most offshore machines are either British, French or German. The trap is that the HALF-PENNY chute will take an American or Canadian dime, and the 1d will take a half-dollar. But if it doesn't say 10¢ or 50¢ (and very few 50¢ American machines were ever made; only 3-reel slots) the machine is most likely British. People miss these clues all the time in their enthusiasm to make a find and a valuable score. Some day these machines will be highly collectible (they finally are in their own countries), but on the North American side of the Atlantic that day hasn't dawned yet. So if it's Un-American it's usually worth considerably less—-3 or 4 times less—-than a comparable American machine. And beware, it can't be fixed. So if you do your research, all of the other "Re's" will fall into place and you'll be well grounded in slot collecting. And you probably won't make many mistakes.

Paupa & Hochriem COLUMBIA, 1897. Gustave Hochriem brought the automatic payout color wheel "Schultze Machine" idea to Chicago, and with partner Joseph Paupa created the 5-way counter wheel machine. Various models were introduced in 1897. The symbol-and-color selective 5-way coin chute is built into the cabinet, upper right. Every color was a winner, as long as you picked it. $3,220.

In addition to the "Six Re's" (real, restored, revamped, replica, (un)realistic and research) you should be aware of the six U.S. and Canadian coins: penny, nickel, dime, quarter, half dollar and silver (or whatever they are made of these days) dollar. Can a coin be significant in the desirability and value of a slot machine? Absolutely. You'll be surprised. To make a fast point, the most desirable coin for a slot machine is the American quarter; two bits, 25¢. Why? Because it's American, it's husky, has value, and is available in great numbers. Taking a close look at these points will be a revelation for some, yet to the died-in-the-wool coin machine collectors they all make sense. There are very valid reasons why this one coin is preferred over any other. First, it's American origins. That means that the machines it will operate are American-made, currently the most preferred slots for both collectors and game room enthusiasts. It also means that the mechanism is repairable machinery, probably made by Mills, Jennings, Pace, Watling or Bally, with spare parts available in quantity (or reproduced to replace the originals) so that maintenance or repair of the mechanism is not a major problem.

Next is the fact that the quarter is husky. This is probably the most amazing part of the preference story. We have to go back in time for the explanation. The very early payout slot machines used coins to trip a lever, at which point a pre-wound clockwork mechanism did things or moved wheels and made the payouts. But the American public didn't accept that. If you swatted or kicked the machine, and it dropped it's payout coins, so much the better. So the clockwork mechanism machines were soon off the market. They ran just fine in other parts of the world where there was more respect for commercial property. But not in the Good Old U. S. of A.

used the inserted coin as part of the mechanism to act as a positive mechanical feature. I'll explain. You drop a coin in one of these older mechanical slot and you see it in a window, then pull the handle on the side. A slide or probe inside the mechanism pushes against the coin and makes the machine operate. What has happened is that the coin itself became a mechanical block, with a part of the mechanism pushing it forward. If the coin wasn't there, the slide would have nothing the push, so no operation.

Later electromechanical and solid state machines, such as the post-1963 Bally and IGT slots, got away from that, and do everything electrically (from "accepting" the proper coin to tripping the mechanism). But the older mechanical slots from 1900 through 1970 all utilize the mechanical block. It's interesting if you watch the process, and follow the stresses place on a coin in an old machine. The fact that the quarter is such a big piece of solid metal makes it a wonderful mechanical block and mechanism mover. A nickel is also a strong, durable coin and is the second choice for preferred coinage. It also makes a good mechanical block, but not quite as good as a quarter.

M.B.M. (Mills) THE KLONDYKE, 1897. Mortimer B. Mills, making cigar venders at his MBM Manufacturing Company in Chicago, added slot machines in the Schultze and Paupa pattern, putting the 5-way color selective coin head on top. The machine was named after the contemporary Yukon gold strike along the Klondyke river near the Alaskan and Canadian border, the last great gold rush. $3,720.

Americans are very hard on their public amusements, and when money is involved, watch out. So other methods of motivation were tried. Cranks that pulled chains or leather belts and released them with a coin were quickly broken. Knobs got knocked off. Coin slides got bent. It was Charlie Fey, the inventor of the 3-reel Bell machine, and some of the other early coin-op pioneers, that figured out how to beat the American penchant for destruction as well as make the machine work better. They

Cowper OREGON, 1898. The first of the color wheel floor machines were electrical, operated by batteries (that ran down!). Invented in Kalamazoo, Michigan, they were produced in Chicago by the Paul E. Berger Manufacturing Company and by the Cowper Manufacturing Company. The OREGON was a later model, with an 8-way coin head. Pick a color, put in the nickel, and push the handle down and hold it. Payout cup at bottom left. $7,020.

Then there is the factor of value. A quarter is still 5X the value of a nickel. Few, if any, other nations in the world offer the same enduring assurance of value as United States coinage. And playing a quarter also feels more risky, because it is a larger denomination coin than a penny or a nickel. Somehow that makes the play more fun. Finally, there's the availability factor. All you need to do to play a quarter slot machine is have some quarters. And if you don't have any, you can go to the bank and get some. United States coins are like U.S. stamps, they have never lost their form or function. A slot machine made to operate on a nickel or quarter in 1898 will operate on today's nickels and quarters. The dimensions are practically the same, so they still work. I am not aware of another country that has the same coinage dimensions they had a century ago. Only in America.

Okay, you say, how come the nickel, or the half dollar, aren't the primary coins of choice? Simply, they aren't as much fun. A slot machine fact of life is that more machines were made for nickel play than any other coinage. The nickel was a major coin from the turn of the century until the 1970s, having a value of 30X a century ago (or the equivalent of $1.50 in today's money) and 15X fifty years ago (or 75¢). That was a pretty big bet then, so most slot machines made between 1890 and 1970 were nickel slots. But coinage has degraded so badly since then that many casinos took out their nickel machines years ago. It's hard to find one in Atlantic City, Las Vegas, on a river boat or an Indian Reservation, and you'll wait in line. A few are stashed around to catch the small time players, but it's the quarter, half and dollar machines that get the play because they're in the majority. Nickels are too much work, and you don't "win" much (I say "win" in quotes, because in a collectible slot no one wins. They just play.). Half dollars are too hard to find, and keep. How many have you gotten in change in the last year? And silver dollars? Forget it. Who wants to tie up a fortune in a toy? And if someone pockets them at a party, where are you going to get more? Not only that, the halves and silver dollars are almost too large for the coin operated machinery to manage. They get wobbly and slightly floppy when they are pushed as a mechanical block, so they aren't preferred.

Notice I haven't mention dimes, or pennies. The early slot makers hated the dime, because it's so thin. A worn dime in an old mechanical slot machine will quite often lap over the coin before or after, and jam the mechanism. Dimes are just too damn touchy to be enjoyable, and very few machines were made for 10¢ play for that reason. They are the least preferred. It wasn't until slots dropped the mechanical block idea and went all electrical that dimes finally found favor, for then they were no more difficult to handle than any other coin. But in the old mechanicals, they're trouble. As for penny machines, they were rarely made except in the Depression years. Between 1930 and 1934 every slot maker produced penny machines, but after 1935 they were a rarity because it was difficult to make money operating penny slots. And even today, there's no feeling of accomplishment when you hit a penny jackpot. So what's 40¢ these days? Not much. Yet some collectors prefer penny machines because of their dating, with anything from the days of the Great Depression having singular historical value.

The rarity of a slot, particularly in the case of significant and highly desirable vintage machines, is never affected by the coinage. But if it's just a gameroom slot, and you're not into the aesthetics of collectibility and just want to buy something to play around with, go for the quarter or nickel machines if you want to play them a lot, or the halves and dollar machines if you want to spice up the play with a little more value. Friends at a party will always remark when they see a dollar slot (or even a half for that matter) because they haven't seen the coins in years. If you're selling, unless you've got a major collectible where it doesn't matter what its coinage may be, be aware that penny and dime machines will probably bring a bit less than the nickel or quarter machines. And you need a special buyer for a half dollar or dollar slot. But on the up side, no matter what you have, if it's fun and it's yours, what does it matter what the coinage is? And one of the most exciting ways to collect slots is to have a whole string of the same machine in a variety of coins, which has a way of making the undoubtedly rare dime and penny models highly desirable.

The differences between an electromechanical and an all-mechanical slot are stunning, in size, weight and performance. With a Bally it's like having a modern casino in the house, with whistles, bells, lights and high hopper payouts, whereas an old mechanical clunks along and spits out a few coins from its payout tube. The difference in cost is stunning, too. Where most old mechanical slots from the 1930s, '40s and '50s sell in the $1,000-$2,000 range (with the Jennings console lightups going for a lot more), the 1963 and later Ballys go for $1,800 up to the $3,000 range. It's looney. They aren't even old, and their supply is virtually unlimited due to casino dumping. Their other

Mills Novelty OWL, 1898. Mortimer Mills sold his company to his son Herbert Stephen Mills in 1897 as his new all mechanical OWL floor machine was being readied for production. H. S. Mills renamed the firm the Mills Novelty Company just as the OWL became the most successful slot machine made to date because it was reliable. Mills adopted the owl symbol as his logotype, remaining in use for the next three quarters of a century. $6,010.

counterparts such as Seeburgs, Mills, Jennings and other electromechanicals don't sell for nearly as much (they generally aren't as flashy or as much fun as a Bally, and the parts are harder to get) yet they still command prices up there with the vintage slots. It will take a few years for all this to settle out, with rarity, age and condition once again taking hold. But for now, it is the heavyweight casino "gameroom" machines that are swamping the collectibles field, so you should be aware. Be aware also of the imports. In the '50s and '60s the British market went full blast, allowing casino gambling starting in 1962. Machines from all over the world congregated there; Mills and Jennings slots from Nevada, and the newer Bally electromechanicals. Plus the British Jubilee, Australian Ainsworth ARISTOCRAT machines, and Segas from Japan. They were dumped by their pub operators as the next generation of the Ballys came on stream. As a result, starting around 1975 and throughout the 1980s, hundreds of these British coinage machines reached American shores, and they're still coming. Like the later Bally slots, they are great as game room machines once they are converted to American coinage, but they are not generally regarded as collectibles. As a result their values are a lot lower than the old mechanicals or the later electromechanicals.

Quick recap. There are generally three groups of automatic payout slots to pick from. The old mechanicals (including the rarities that stretch back to the 1890s) have been the collectibles of choice from the 1960s to today. They are what most collectors prefer, but every class of machine has its place. The imports came next, and have not been classified or priced as collectibles. Finally, the Bally electromechanicals and the even later Sircoma, IGT, Universal and other modern slots and payout blackjack and poker machines have started to carve out a place as the collectibles of the future. One of their advantages is that the coinage doesn't matter as all coins work equally well. Depending on what you like—-age and patina, just a slot to play, or a game room extravaganza—-the mechanicals, imports, Bally and other machines are all valid slots, with specific price levels to match.

Western Novelty HONEST JOHN, 1899. Large automatic color wheel floor machines became the rage in saloons and gambling locations, with the makers proliferating. The Western Novelty Company was in Kalamazoo, Michigan, an early hotbed of slot machine development, at a time when anything west of the Allegheny mountains was called "Western." This is the 8-way HONEST JOHN with an enormous coin head. It was also made in a 6-way model. $10,100.

Mills Novelty JUDGE, 1899. An advanced development of the OWL, using the same cabinet, the Mills JUDGE added a color viewing window below the coin head, replacing an instruction panel. A color flag flipped in place when a color was selected by playing a coin, eliminating disputes at payout time in the event of a winner, or a claimed winner that wasn't valid. $6,025.

Mills Novelty DUPLEX, 1899. One of the few vintage slot machines that defines a significant antique slot machine collection. The DUPLEX combines the pocket machine format of the bartop Clawson THREE JACKPOT in a very large floor machine with ten much richer pots, with payouts based on a wheel spin. Two 5-way selective color coin heads (making it a 10-way machine) and twin payout cups give the DUPLEX its name. Patterned after an earlier Mortimer Mills machine called the TURTLE. $30,100.

McDonald BANNER, 1900. McDonald had worked with Mortimer B. Mills before H. S. Mills took over the company, after which he opened his own shop in Chicago. BANNER was a 6-way color wheel with the same color flag feature as the Mills JUDGE. It came plain, and as the MUSICAL BANNER shown here, in which a Swiss music box played a contemporary air when a coin was dropped, giving you something for your money to challenge the anti-gambling laws. $8,520.

Mills Novelty DEWEY, 1899. Major media figures had little protection against exploitation in the 1890s. So Mills Novelty appropriated the fame of Spanish-American War naval leader Admiral George Dewey, the "Hero of Manila," and put his name and face on the front of a slot machine. The DEWEY became the most famous Mills floor machine, and was still being sold as late as the early 1930s. $9,020.

Mills Novelty BROWNIE, 1900. The Palmer Cox "Brownies" were the best known cartoon figures in the country (about as well known as today's The Simpsons) at the turn-of-the-century. So it wasn't surprising they were featured in a slot machine, with their images in the top coin head casting (with probably no royalty to Cox). The Mills BROWNIE utilized a new countertop size wheel mechanism. $3,020.

Automatic Machine GABEL'S TWO-BITS DEWEY, 1902. John Gabel wasn't above copying the competition, making a virtual duplicate of the Mills Novelty 6-way DEWEY as did others. The glass shows the symbol for a horse bit and a 2, meaning "two bits," or 25¢ play. Earlier Automatic Machine models just had the machine name. Later models added Gabel's name to the glass so operators knew where it came from. A 7-way model was made called GABEL'S LEADER. $10,100.

Caille Bros. TWO-BITS, 1902. The incredible good looks of the Caille Bros. floor machines gave the competition a solid run for its money, making the Detroit firm the number two supplier of these gambling machines, after Mills Novelty. A variety of trims and finishes were offered. This TWO-BITS model (meaning it was 25¢ play) was offered with nickel plate or anodized copper castings, with both mixed in this model, with a cabinet in stained wood or a marbleized finish, the latter as shown. $10,100.

Caille Bros. NEW CENTURY DETROIT, 1901. Mills Novelty named the visible payout mechanism backup model of their famous DEWEY the CHICAGO after its home city, so Caille Bros. did the same thing with their equally famous PUCK, naming it the NEW CENTURY DETROIT after its producing city. The visible plungers coming through the holes in the disc and the half-ringed pointer at the top of the wheel meant it couldn't be bugged to cheat the players by making it impossible to hit a major payout. Only it could. $10,100.

Automatic Machine GABEL'S NIAGARA, 1902. Mills Novelty alone didn't make Chicago the "Coin Machine Capital of the World," but a cluster of surrounding manufacturers did. The Automatic Machine & Tool Company, under the management of John Gabel, was one of many local floor machine producers making original machines. NIAGARA was a 6-way competitor to the Mills DEWEY, available with or without music. $12,100.

Victor Novelty VICTOR NATIONAL, 1904. Yet another Chicago floor machine maker, the Victor Novelty Works was founded in 1904 to cash in on the slot machine boom. Machines were similar to Mills and Watling, with the NATIONAL a 6-way competitor to the Mills DEWEY, with or without music. A variety of models were made, including VICTOR, JACKPOT, VICTOR 4-BITS and others. Later models added the VICTOR name to the glass in Gabel fashion. $10,100.

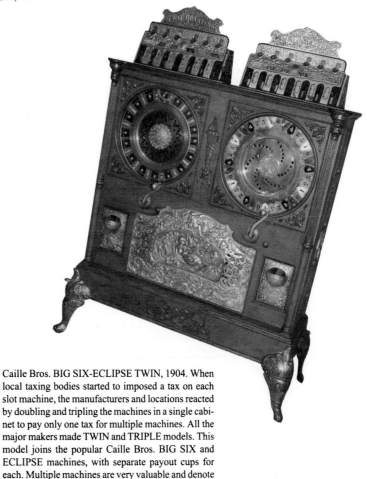

Mills Novelty ROULETTE, 1901. One of the most exciting of the mechanical payout slot machines, the Mills ROULETTE used the mechanism of an earlier machine called the 20TH CENTURY turned upright so the disc was on the top. In its place was a cast roulette wheel with cups to hold the ball when it stopped, with the mechanism reading all plays and making appropriate payouts. $35,100.

Caille Bros. BIG SIX-ECLIPSE TWIN, 1904. When local taxing bodies started to imposed a tax on each slot machine, the manufacturers and locations reacted by doubling and tripling the machines in a single cabinet to pay only one tax for multiple machines. All the major makers made TWIN and TRIPLE models. This model joins the popular Caille Bros. BIG SIX and ECLIPSE machines, with separate payout cups for each. Multiple machines are very valuable and denote major collections. $35,100.

Mills Novelty CRICKET, 1904. Regarded by many as the most entertaining slot machine ever built, the CRICKET is an overgrown "Jacks" machine with nine jackpot pockets. You put the coin in the head, register it with a lever, and hit the shooter. Your coin leaps across the playfield, bouncing over the pins to land a loser, or in a pocket slot, enabling you to dump the pot into the payout cup. Caille Bros. made a virtual duplicate called BULLFROG. $24,050.

Caille Bros. PEERLESS, 1907. Following the Mills Novelty ROULETTE, Caille Bros. made a competitive 7-way 80-pocket machine in 1904 also called the ROULETTE. In 1907 the machine was improved and changed to a 60-pocket version as the PEERLESS, with mechanical improvements to work better. This is one of the most sought after antique slot machines, with less than a dozen known. It is said that ownership of any Caille roulette machine makes the collector world class. $45,100.

Caille Bros. CENTAUR, 1907. The arrival of the countertop cast iron payouts didn't deter the floor machines. At first. The Caille Bros. CENTAUR was essentially a jackpot version of the ECLIPSE, made to compete against the Mills DEWEY JACKPOT. But it is light years ahead of the Mills product in appearance, from its massive coin head to its flashy punched disc. The CENTAUR came plain, and with music. It was a popular seller until World War I. $11,100.

Mills Novelty CHECK BOY, 1907. The newest entry in the heavyweight countertop payout class was the Mills 6-way CHECK BOY, a development of the Paupa & Hochriem single reel ELK and PILOT formats. Mills called it a "Miniature DEWEY" because it made as much money as the floor machine. Its good looks and performance led to copies by other manufacturers, with Caille Bros. making a virtual duplicate. The payout cup is on the right side, behind and below the pull handle. $4,220.

Watling LITTLE SIX, 1909. Even the massive floor color wheel format was miniaturized to iron countertop machines. Watling created a line of nickel plated cast iron cabinet 6-way dial machines called the CHECK BOY (picking up the Mills Novelty machine name), STIMULATOR and LITTLE SIX, the latter the same as the STIMULATOR with a different dial. Picking any winning color pays 5 for 1, giving the house the same advantage as a casino Big Six wheel. $5,020.

Watling EXCHANGE, 1909. An original Watling idea. Taking the basic mechanism of THE JEFFERSON, the cash payouts were eliminated to be replaced with trade tokens. That enabled them to say "No Gambling" on the front glass while the machine kept the money. What the player got if they won was trade checks valued from 10¢ to $1, turned over to the proprietor for an exchange in goods or services. This was a store or cigar counter machine, not for saloons. 2,820.

Watling THE JEFFERSON, 1909. A big name in Chicago slot machine circles, and locally second only to Mills, the Watling Manufacturing Company was formed in January 1902 out of the earlier D. N. Schall & Company, with brothers Tom and John Watling taking over the older firm. Most early production duplicated other machines, with a few original models showing up. THE JEFFERSON was a counter payout wheel in the same class as the Mills BROWNIE, only smaller. $3,010.

Caille Bros. LIBERTY BELL GUM VENDER, 1911. The race was one. The leading makers copied Mills, bringing out their own LIBERTY BELL machines with Caille Bros. and Watling making practically identical models. Then they started making alterations. Caille added an enormous gum vender on the top of their machine as the LIBERTY BELL GUM VENDER. Play the coin, push the plunger and you got a 5¢ package of gum with every play. This is a very rare machine, with only one known. $11,050.

Eagle EAGLE, 1910. Another variation of the cast iron countertop 6-way color wheel theme. The EAGLE was made by the Eagle Manufacturing Company of Chicago, a reformed and renamed Victor Novelty Works, the maker of the older floor machines. The EAGLE came in straight payout, and as the EAGLE GUM VENDER with the color wheel covered. The payout cup is on the front, lower left, and large enough to hold the 20 coins paid on blue, with only one blue spot on the 24 stop wheel. $6,020.

Mills Novelty LIBERTY BELL, 1907. Mills Novelty re-engineered and duplicated the 3-reel LIBERTY BELL after Charlie Fey brought it to them for volume production. They encased the machine in a heavy nickel plated iron cabinet. Very heavy! Nicknamed "Iron Case," with all the meaning that entailed. The machine name and reel symbols were retained, as well as the toed "Tiger Feet." A lot of advertising fanfare accompanied its introduction. Very desirable collectible. $9,010.

Watling LONE STAR SPECIAL, 1910. The day of the floor machine was far from over by 1910, with newer developments still coming. The Watling LONE STAR SPECIAL showed off its mechanical guts to any doubter of its integrity. It also vended packages of gum with every play, dropping the product in the wide tray just below the payout cup, the "Special" part of the name. Play is by a nickel plated spinning pointer rather than a spinning wheel. $12,050.

Watling DERBY, 1911. Developed by Gustav F. W. Schultz in San Francisco in 1906, the DERBY was re-engineered improved and enlarged by the Watling Manufacturing Company, entering production in Chicago in 1911. This race game runs small horses with distinctively marked jockeys in a circle and pays off on the winning colors. Two are known, both four as basket cases with many parts missing, to be restored working condition at great expense by their owners. $85,10

Mills Novelty LIBERTY BELL, 1910. The wave of the future. After the stunning success of the original Mills LIBERTY BELL, the machine was further refined. By 1910 the mechanism and case had been improved and the feet were in a sturdier Arts-And-Crafts movement styling. The OPERATOR BELL model (after the operators that placed them in locations, such as saloons, ice cream parlors or stores), replaced the Fey graphics with the familiar Mills fruit symbols. This is the machine that the rest of the industry blatantly copied. $7,510.

Watling BIG SIX O. K. GUM VENDER, 1911. Watling stuck their half-circle O. K. feature on top of their version of the BIG SIX and added their own side gum vender to one-up the Caille/Fey model of the same machine. The castings differ, with the Watling name over the 6-way coin entry, a chrysanthemum symbol on the lower front where the Fey logo was, and a "6" in the star above to replace the Fey initial, with exotic fruit symbols on the reel. Otherwise it's nuts on. $10,050.

Mills Novelty MODEL NO. 7 COUNTER O. K. GUM VENDER, 1911. Bell machine development was rapid, and varied. For its 7th version Mills created a wooden cabinet with a deeper base to hold a larger cashbox, and added a patented O. K. feature. This extra dial in the award card indicated the number of tokens won, except you had to play another nickel to get them. So you always knew what you'd get each time you played, and no longer gambled. And the location got two coins. $6,010.

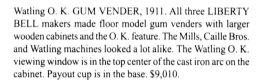 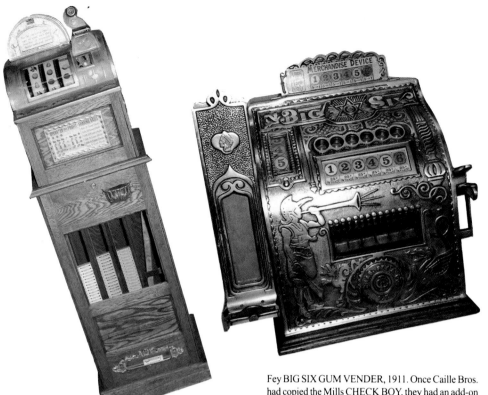 

Watling O. K. GUM VENDER, 1911. All three LIBERTY BELL makers made floor model gum venders with larger wooden cabinets and the O. K. feature. The Mills, Caille Bros. and Watling machines looked a lot alike. The Watling O. K. viewing window is in the top center of the cast iron arc on the cabinet. Payout cup is in the base. $9,010.

Fey BIG SIX GUM VENDER, 1911. Once Caille Bros. had copied the Mills CHECK BOY, they had an add-on development in the wings. By this time Charlie Fey's relationship with Mills had soured, so he began cooperating with Adolph Caille in Detroit. This led to the Fey BIG SIX, with a re-cast Fey front and a new color spot reel on a Caille Bros. CHECK BOY mechanism. Fey also added his simplified GUM ATTACHMENT to make it a gum vender, with payouts in trade checks. $10,050.

Caille Bros. SILVER CUP, 1912. The close cooperation between Charlie Fey in San Francisco and The Caille Bros. Co. in Detroit is nowhere better revealed than in the case of the SILVER CUP. Fey created the machine in 1910, with a somewhat heavyhanded 5-way coin head and a payout cup under the right side of the front. Caille made a refined version in 1911 as the ORIGINAL SILVER CUP, and then further refined the machine into this 1912 version with the payout cup in the center. $9,010.

Mills Novelty SILVER CUP, 1914. Not to be outdone, Mills Novelty stuck two spinning discs on top of their CHECK BOY to create a SILVER CUP model. That seems to have been dropped, for in 1914 a modified Mills BROWNIE split into a double-dial machine was advertised as the SILVER CUP. The front casting is highly imaginative. The two dials become large eyes in a symbolization of the Mills owl logotype. When first found this machine was thought to be a prototype. Others have been found since. $7,510.

Watling O. K. DELUXE NO. 25, 1914. By 1912 Caille Bros. and Watling had refined their LIBERTY BELL and OPERATOR BELL machines so that they no longer looked like the Mills originals, with both makers calling their machines DELUXE models. The new streamlining showed up in a variety of guises, with the Watling O. K. DELUXE NO. 25 the top of the line in 1914. Even the O. K. feature has been improved. Castings differ slightly from the similar NO. 24 model, with "Chicago, Ill." smaller. $12,050.

Mills Novelty 1915 OPERATOR BELL, 1915. This is the model that set the pattern for Bell machine styling for the next 15 years. Mills took their "Iron Case" OPERATOR BELL, removed the feet, put a large cash box in its place, and extended the cast iron front full depth in an oak cabinet. It took about 20 pounds out of the machine. The bell symbol at the bottom left carried the model date: 1776 to 1915. These changing dates remained as model designators for another six years. $1,810.

Jennings OPERATORS BELL, 1922. The company that would be the major competitor to Mills started out in Chicago in 1907 as the Industry Novelty Company, changing its name to O. D. Jennings & Company after WWI. They revamped Mills mechanisms in wooden cabinets, and soon were making their own machines from the bottom up. This is their original styling from 1916 through 1925, with a "curved glass" viewing window. Its nickname was "Gooseneck." $1,210.

Mills Novelty 1922 OPERATOR BELL, 1922. The next big change in Mills slots. The 1922 model had a new mechanism and an aluminum front, the wonder metal coming out of World War I. As a result even more weight was taken out of the machine. Note the missing model date in the bell at lower left, although the "1776" remains. This version was called the "Bell" model because of the bell symbol on the front. $1,210.

Mills Novelty/Iowa Novelty O. K. MINT VENDER, 1923. In 1923 the appearance of the Mills Bell line was changed once again with a new cast aluminum front that perched two owl heads on top of design columns. An early customer for an exotic model was the Iowa Novelty Company in Cedar Rapids, Iowa, who ordered private label side vender versions with complete aluminum cabinets, replacing the wood on the sides and base. These are rare machines. $2,710.

Caille Bros. O. K. FRONT VENDER, 1923. The Caille post-WWI model of the Bell machine had a center pull handle, with a big "U" shaped bracket inside to reposition the arm. In its OPERATOR BELL model it was called the "Victory," after the Allied success in the war. The 2-column vending front mint vender version was announced in 1923, and included the O. K. feature. It was called the "Victory Mint Vender." Early fronts were cast iron, later aluminum. $2,210.

Jennings AUTOMATIC COUNTER MINT VENDER, 1924. The classic Jennings wooden cabinet styling in its side vender model. As the AUTOMATIC COUNTER VENDER 1919-1923 it vended gum. In 1924 this was altered to mint vending, resulting in a much more successful machine for store locations, a need once Prohibition was the law of the land. The O. K. feature is on the cast front plate. Without the side vender, and paying trade tokens, this was called the THE TRADER. $1,410.

Jennings AUTOMATIC MINT VENDER, 1924. A newer member of the Jennings AUTOMATIC COUNTER VENDER line, although it now shows off the mints that can be earned through play in a cast aluminum front. This gave the machine the nickname "Display Front." But that's all they were. When the O. K. feature indicated you would receive so many mints or premium checks on the next pull, you played, and took the returns in candy or cash. It got around anti-gambling laws. $1,520.

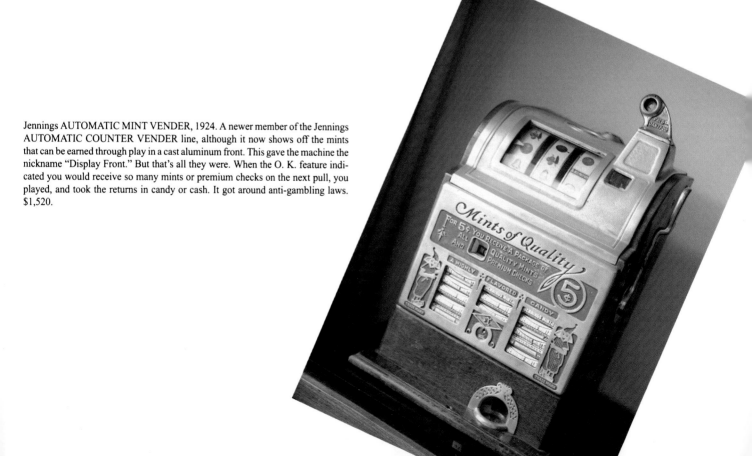

Mills Novelty 1925 OPERATOR BELL, 1925. The classic Mills middle '20s Bell machine, nicknamed the "Owl" model for the two owl heads on the left and right sides of the front. The show window glass is flat, and a small viewing window at the right below the coin chute lets you see if the coin is real and not a slug. You can date Bell machines by their serials; this is 121,140 in 25¢ play, big coinage for the '20s. $1,210.

Jennings TODAY VENDER, 1928. On the Mills and Pace front venders, to get your mints you turned a knob on the side and got what came next. Jennings made a dramatic improvement over that with their selective TODAY front vender of 1926, which allowed you to select the flavor of your choice and pull the knob beneath its column to vend. The advanced "All Quality Confections" model of 1928 picked up the colorful "Dutch Boy" graphics first used on the "Display Front" machine. $1,410.

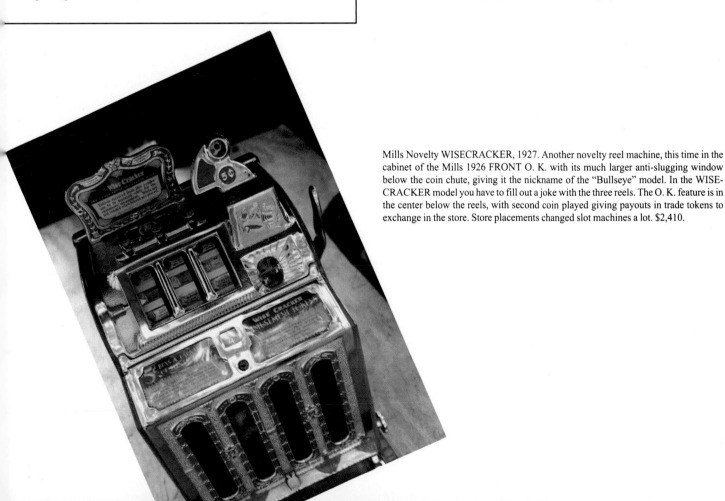

Mills Novelty WISECRACKER, 1927. Another novelty reel machine, this time in the cabinet of the Mills 1926 FRONT O. K. with its much larger anti-slugging window below the coin chute, giving it the nickname of the "Bullseye" model. In the WISE-CRACKER model you have to fill out a joke with the three reels. The O. K. feature is in the center below the reels, with second coin played giving payouts in trade tokens to exchange in the store. Store placements changed slot machines a lot. $2,410.

Skelly 1928 BELL JUNIOR, 1928. Pace may have bought out Skelly's Bell machines, but the Skelly Manufacturing Company of Chicago went on to produce some very strange machines that somehow never caught on, making them rare indeed. The mechanism of the 1928 BELL JUNIOR revolves a single drum with all three symbols in line. It must have been made as a portable machine as it has a carrying handle on the top attached to the gooseneck. A similar model is 1930 POT BOY. $1,810.

Pace 1927 FRONT VENDER, 1927. Ed Pace, a former slot machine salesman, bought out a small company called Skelly that made Bell machines, and put his name on them. The Pace Manufacturing Company of Chicago became a major producer. This is machine serial 599, perhaps the earliest known as Pace serials may have started at 501. The F. O. K. design is classic, copied from Mills, with the circular anti-slugging escalator below the coin chute uniquely Skelly, and then Pace. $1,810.

Caille Bros. O. K. FRONT VENDER, 1924. Novelty reel slot machines are eagerly sought after by collectors as they are rare and unique. This is the "Beachcomber" version of the "Victory Mint Vender." The show window has added a mask of palm trees. The reel symbols are parts of a human body, with playing card spots in the background. Get the right three and you get a payout, with the pirate face symbol shown with the housewife body and legs a two coin winner. Very desirable. $2,515.

Pace 1928 AUTOMATIC SALESMAN, 1928. Matching the Mills REGULAR O. K. "C. O. K." (for COUNTER O. K., the original name of the Mills model) the Pace AUTOMATIC SALESMAN "C. O. K." offered the O. K. feature in a small window at the far left even with the payline. Pace called this feature Future Pay. The Pace fruit symbols also matched those of Mills, although the Bell-Fruit-Gum bar symbol carries the notation "Pace Manfg. Co. Chicago." $2,010.

Pace 1928 FRONT O. K., 1928. The Pace "F. O. K." didn't change much in a year, although the gooseneck casting began to look more like the Pace components of later years. Pace also started coding their serial numbers by the type of machine, with this being F-3,114, with the "F" standing for front vender. $1,810.

Pace 1928 OPERATOR BELL, 1928. Pace matched the Mills Bell line model for model, in distinctive Pace cabinets with the Skelly coin escalator. In keeping with Skelly fashion, the model date was made part of the front casting. This was dropped the following year. Available coinage was 5¢, 10¢ and 25¢. Serial number of this machine is 2,285. $1,510.

Roberts Novelty IMPROVED JACKPOT, 1929. Roberts Novelty was a slot machine distributor in Utica, New York, with a big idea. They made replacement jackpot fronts to upgrade older machines, and in most cases that meant Mills. Mills issued a JACK POT kit, and machine, late in 1928 known as the "Torch." At the end of the year an improved JACKPOT "Poinsettia" was also introduced. Both were often revamped with larger jackpots, such as the Roberts. $1,410.

Caille Bros. SUPERIOR JACKPOT, 1929. The improved barless jackpot front on the Caille SUPERIOR JACKPOT. Twin payout cups are characteristic of the early jackpot models by all of the manufacturers, one for the regular tube and slide payout and the other for the pot. This model became known as the "Design Front" for the elaborate scrollwork in its casting. The fact that the pot jutted forward from the front attracted players. $1,210.

Caille Bros. SUPERIOR JACKPOT, 1928. Caille was the first major producer to introduce a jackpot model, providing JACK-POT revamp kits for their SUPERIOR Bell machine in 1928, followed by factory produced models. The factory versions had a variety of front designs and pots, as if indecision regarding optimum placement was under consideration. Early models had barred windows, later models were open. Operators sometimes sawed off the bars to make their machines appear current. $1,210.

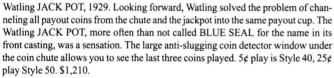

Watling JACK POT, 1929. Looking forward, Watling solved the problem of channeling all payout coins from the chute and the jackpot into the same payout cup. The Watling JACK POT, more often than not called BLUE SEAL for the name in its front casting, was a sensation. The large anti-slugging coin detector window under the coin chute allows you to see the last three coins played. 5¢ play is Style 40, 25¢ play Style 50. $1,210.

Jennings PRIMER JACKPOT VENDER, 1929. Jennings stuck jackpot fronts on its upgraded flat glass OPERATORS BELL machines in 1929, and then proceeded to waffle over its location on the front. The dancing figures were retained, with the machine nickname "Blue Boy." The original small pot added a reserve inside to refill the pot when hit, giving the newer model the name PRIMER (which some dealers corrupted to "Premier")."All Quality Mints" side vender. $1,610.

Jennings BASEBALL, 1930. The Jennings TODAY was chosen for modification for the latest fad: baseball. Radio had brought it into the American home. A new interactive front, baseball reels, and a complicated instruction panel created a new game, and the Jennings BASEBALL took its place with the Mills and Watling models made a year earlier. Reel stopping pushbuttons made it a skill machine. $2,715.

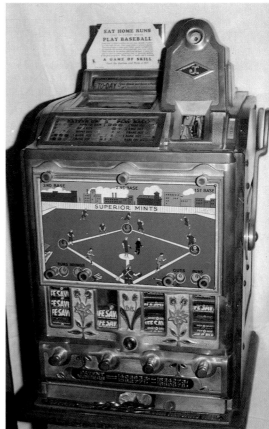

Watling 1930 BROWNIE JACKPOT, 1929. Watling looked backward and forward at the same time to get whatever sales they could. Once the jackpot was recognized as a hot item, they took their old IMPROVED BROWNIE and earlier BROWNIE JACKPOT patterns and dies and introduced an upgraded 6-way counter wheel payout with its own jackpot front, a small window at the far left. These machines may look a lot older, but were introduced in 1929 and sold until 1935. $2,015.

Jennings PERFECTED JACKPOT, 1930. The wandering jackpot finally found a place to rest on the Jennings front, and the dancing "Blue Boy" symbols took on a variety of colors. The factory painted the machines to order. Deco was beginning to appear as a design element, as shown on the sides. Jennings still had two payout cups. The pot placement allowed a reserve behind it. $1,410.

Pace BANTAM JAK-POT, 1930. A refined case and improved jackpot window brought the BANTAM JAK-POT into the '30s, and the beginning of a Golden Age of slot machines. The next decade would provide more new ideas and innovations in slot machines than any time frame before or since. It took 80 to 90 plays for the Pace to refill its "Jak-Pot", reason enough for machine makers to consider adding reserve pots for an instant refill to stimulate play. $1,210.

A. B. T. TRIP-L-JAX, 1930. A. B. T. was a Chicago pinball and counter game maker that got into "Jacks" machines with their TRIP-L-JACKS, and made a killing. Put the coin in the right side, whack it with your finger, and it jumps across the pins. If it lands in a pot chute, turn the crank and the pot below it dumps. In effect it was a three jackpot machine just as jackpots were getting introduced. Everything old was new again. $860.

Pace BANTAM JAK-POT, 1929. A Skelly design, the Pace BANTAM achieved its smaller size by eliminating the bottom cash box, replacing it with a coin bag. Straight Bell BANTAM models started in 1928, with the Jak-Pot (the Pace name for jackpot) added in 1929. 25¢ play is very big money for the day. The handy Skelly carrying handle remains, to be quickly dropped in later models. Serial is BJ3,282, with "BJ" signifying BANTAM JAK-POT. $1,210.

Pace BANTAM EVER-FULL JAK-POT, 1930. Pace added the play maintaining reserve jackpot to its 1931 BANTAM model late in the year. The reserve refill pot was held in the small barred windows above the paying jackpot in the center, with decorations in the casting below the pot. The reserve refilled the pot on the next play after a jackpot was hit. Pace retained two payout cups. $1,210.

Pace 3 JACKS, 1930. Even the Pace Manufacturing Company got into the "Jacks" act and made three jackpot machines under their own name and under private label for the American Sales Company in Green Bay, Wisconsin. Pace called their version 3 JACKS while American called theirs TRIP-L-JACKS, dropping the name when it conflicted with A. B. T. These machines are fast and a lot of fun to play. They are also relatively inexpensive as collectible payout machines. $760.

Jennings ROCKAWAY, 1931. Another old idea brought back to life by Charlie Fey and made by Jennings was the five jackpot ROCKAWAY, a modernized version of an old 1890s game called either the INVESTOR or the BONANZA made by Little Casino in Rochester, New York; Charles T. Maley in Cincinnati and others. The coin jumps around on a spinning disc in the playfield, and if it lands in a pot slot the jackpot drops automatically, no knob to turn. It didn't do well. $1,010.

Field 4 JACKS, 1930. A small punchboard producer in Peoria, Illinois, named the Field Paper Products Company undertook the marketing of a modernized aluminum version of the old Clawson THREE JACKPOT of 1893 with four pots as the Field 4 JACKS. It was such a major success Field found itself in an oversold situation, and changed its name to the Field Manufacturing Corporation. It wasn't long before half a dozen others made the machines, but there was room for everybody. $660.

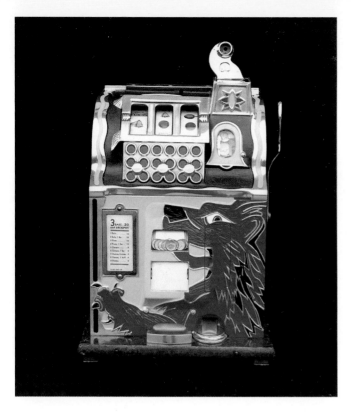

Mills Novelty SILENT GOOSENECK, 1931. Some operators were afraid of new fangled developments, like the escalator. So Mills created a SILENT GOOSENECK model that had all of the outside dimensions of the old gooseneck machines, but looked entirely different due to the graphics design of the painted cast aluminum top and front. The suggested machine nickname was "Lion's Head," but most often the operators called it the "Lion Gooseneck." $1,410.

Watling TWIN JACKPOT, 1931. Classically styled, the Watling TWIN JACKPOT faced two jackpots to the players. There was always another one to hit if one paid, eliminating the need for a reserve as it was built-in. This is the initial "Small Jackpots" version. A year later they were greatly enlarged to create the "Large Jackpots" model. This format was so popular, Watling retained it for another quarter century in their ROL-A-TOP line. Operator nickname is "Fancy Front." $1,410.

Mills Novelty SILENT, 1931. The Bell that revolutionized slot machine design and appearance, and led to the '30s Golden Age. The Mills mechanism was completely redesigned to eliminate noise and wear spots. The cabinet was the first to use the front as a graphics substrate, with a right facing "War Eagle" commanding the design and its two jackpots. A major addition was the escalator, a horizontal track that showed the last five coins played. Highly desirable collectible. $1,660.

Benton Harbor NEW IMP, 1931. The challenge to create a new, extra small payout Bell slot machine was joined by many manufacturers, in Chicago and elsewhere. The Benton Harbor Novelty Company was located in Benton Harbor, Michigan, a highly industrialized automotive feeder city a little over an hour's drive from Chicago. The IMP came out in the summer of 1931, with the NEW IMP jackpot added in the fall. The machine was redesigned in 1932 to add a bugle blower on the front. $910.

Vendet MIDGET, 1932. Vendet Manufacturing Company of Chicago made a small fruit symbol Bell by cutting it down to two reels. A second version has dice symbols. It came in both Bell and side vender models. It didn't fare well, with the whole line and its tooling sold out to the Superior Confection Company of Columbus, Ohio, who sold it in 1933 to the Burtmeier Manufacturing Corporation back in Chicago, who redesigned and brought it back as the PONY. $1,260.

Mills Novelty SILENT JACKPOT FRONT VENDER, 1931. Even the venerable F. O. K. came in for its SILENT redesign and stylized front. The SILENT JACKPOT FRONT VENDER was also called the SILENT F. O. K. as well as "Modern Front." The eagle symbolism of the SILENT "War Eagle" payout Bell was doubled with eagle heads both left and right on the front. This was also the first Mills machine to get away from the two cup payout, using a wide trough in the Watling fashion. $1,660.

Jennings RAINBOW, 1932. Jennings kept trying to come up with old machine ideas that might find a new audience. RAINBOW is essentially a five jacks machine utilizing a cabinet similar to their earlier ROCKAWAY, minus the spinning disc on the playfield. By the end of 1932 the "Jacks" market had stabilized, and newer and more reliable machines had replaced the slapdash products of 1931. Never a mainline slot, they long retained a loyal audience and market. $1,010.

Rock-Ola FOUR ACES BALL GUM VENDER, 1932. Rock-Ola entered the "Jacks" market in 1930, and quickly moved itself into a leadership position with the small machines. FOUR ACES is their peak model, glistening in an all aluminum cabinet. It came in both plain and ball gum vender models, with the ball gum window and dispenser centered between the pots and the playfield. $910.

Caille Bros. SUPERIOR GRAND PRIZE, 1931. Caille was being severely outclassed by the redesign throughout the industry, and countered with a fourth reel model called the SUPERIOR GRAND PRIZE just about the time the company changed ownership hands. The fourth reel sets up an additional award added to the regular payout. The jackpot front is the larger "Reserve" model that has a second pot in reserve, instantly refilling a hit pot on the next play. This machine is fairly rare. $1,615.

C & F BABY GRAND, 1932. Yet another solution to the miniaturized Bell. The C & F Manufacturing Company was located in Chicago, with strong connections to the Field Manufacturing Corporation in Peoria, possible the "F" in C & F. Some BABY GRAND machines even have C & F name tags with a Peoria address. BABY GRAND is a full payout Bell machine, packed into a small cubed cabinet. It weighs a wispy 27 pounds, one-third to one-quarter the weight of a full size slot machine. $1,010.

Rock-Ola MILLS FRONT O. K. JACKPOT, 1932. Rock-Ola had a substantial business in replacement jackpot fronts to upgrade old OPERATOR BELL and non-reserve jackpot machines, running first in the industry with Pace coming in second. Because Mills Novelty machines were the most numerous, Mills jackpot fronts were generally the line leaders, although fronts were also made for Jennings and Watling machines. Pace made their own. This front added a jackpot to the F. O. K. $1,410.

Watling TWIN JACKPOT BALL GUM VENDER, 1932. The Great Depression greatly influenced slot machine design. 1932 was at the depth (or the height) of the depression, so machine makers countered with penny play machines. Typical was the conversion of the front vender to a penny gumbal model. The TWIN JACKPOT still had "Small Jackpots," but a major change was coming. Watling also made a TWIN JACKPOT 2-FOR-5 model that gave two handle pulls for a nickel. $1,810.

Jennings VICTORIA SILENT, 1932. The Mills SILENT line put pressure on Jennings, who came out with their own versions soon thereafter. The re-engineering wasn't as extensive, and the sound deadening was by rubber bumpers, but they had competitive machines in a hurry. Both escalator and gooseneck models were produced, just as Mills did, with Jennings using peacocks on the front instead of eagles. This is the VICTORIA SILENT "Peacock Gooseneck" model. $2,210.

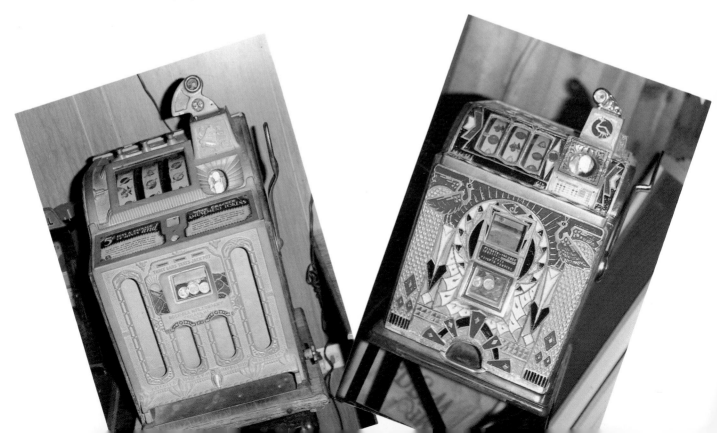

Mills Novelty SILENT GOLDEN, 1932. The 1932 addition to the Mills SILENT escalator line was a machine that introduced a new payout concept in the Gold Award. A small window held special golden colored tokens, worth more than the jackpot and redeemable in cash or trade. Hit the three Gold Award symbols and its yours. The SILENT GOLDEN was one of the most striking cabinet designs ever created by Mills. It received the nickname "Roman Head." $2,010.

Pace 1934 COMET VENDER, 1933. The older looking full size Pace Bell machines came in for their restyling in 1933 as the 1934 COMET line. The TWIN JAK-POT feature was used on most models, including both two column jackpot and four column plain front venders. Pace didn't figure out how to include the rotary escalator, so made the front venders with "Bullseye" anti-slugging viewers. They still looked a lot like Mills machines, but were getting away from that. $1,410.

Pace BANTAM MINT VENDER, 1933. The "Bullseye" used on the 1934 COMET had been worked out for the BANTAM, by now reaching the end of its development line. The ultimate "Christmas Tree" model was the 1933 BANTAM MINT VENDER with a display front, side vender, jackpot with reserve and "Bullseye" window. Pace called their BANTAM reserve jackpot EVER-FULL. $1,410.

Mills Novelty MYSTERY, 1933. The big Mills Novelty SILENT introduction for 1933 was a machine that changed the course of slot machine payouts. Standard Bell payouts had been 2-4-8-12-16-20. But when the MYSTERY was put out it had an award card that said it would pay 2/4 and 4/6 on the lower wins, while actually paying 3 and 5. People thought the machine was overpaying because it was broken, and it became an instant hit. Later models have metal award card. Its nickname is "Blue Front." $1,210.

Watling 1934 BABY GOLD AWARD VENDER, 1933. Watling also made a smaller Bell machine in the BANTAM class, and they did it the same way as Pace by taking parts out of the machine, heightening the effect by smaller narrower reels. They were displayed in the same show window, but were optically further reduced in size by tapering the sides to make a smaller viewing hole. In the GOLD AWARD model a GA ("Gold Award") token panel is tucked above the "Small Jackpots," plus mints. $1,710.

Mills Novelty SILENT JACKPOT FUTURE PLAY, 1933. Some territories wouldn't allow gambling machines, and others wouldn't even allow venders. To circumvent this Mills developed the FUTURE PLAY INDICATOR, a device not unlike the half-circle Watling O. K. feature put on top of their machines twenty years earlier. This add-on predicted the payout result of the next play, so it wasn't gambling. It could be attached to most Mills Bells, including the "Modern Front." $1,810.

40

Jennings CENTURY TRIPL-JACK, 1933. The first of the truly modern Jennings Bell machines, the CENTURY was named after the Century Of Progress World's Fair in Chicago in 1933. Its swirly Deco design gave it the nickname "Modernistic Front." In Jennings terms it was the VICTORIA MODEL G, but it was a far cry from the VICTORIA design track. It came in both gooseneck and escalator models, the latter shown here. $1,510.

Mills Novelty MYSTERY GOLDEN, 1933. The MYSTERY became the Mills Novelty substrate machine. They hung everything they could on the basic construction to create a wide variety of models. This is the MYSTERY GOLDEN, so named because it added the GA feature, upper front, right, and the appropriate award card. The "Blue Front" was the most numerically produced Mills machine, produced in various models from 1933 until 1942. Early model has framed paper award card. $1,410.

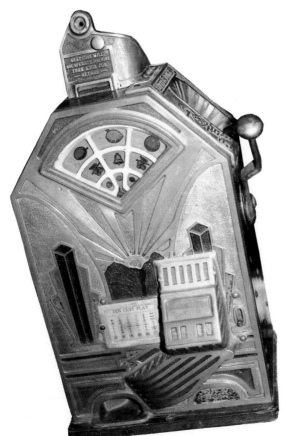

Jennings SELECTIVE LITTLE DUKE JACKPOT, 1933. New and original slot machines were rare in the early 1930s, but Jennings came up with one. The triple disc (not reel) spinning LITTLE DUKE originated in France, with Jennings having North American rights. The French sun and moon symbols were changed to American fruits, and it was spun off into a myriad of models. The "Selective" name came from the late model coin head capable of selecting coins and rejecting slugs. $1,510.

Jennings SELECTIVE LITTLE DUKE TRIPL-JACK VENDER, 1933. The Tripl-Jack was a double jackpot with a third in reserve that Jennings applied to the LITTLE DUKE and full size CENTURY Bell machines. The fact the pots were small didn't seem to deter players as they were very popular. The LITTLE DUKE gum vender is on the left side, uncommon for '30s slot machines. The handle was so tight to the cabinet there was no room on the right side. It also came without the selector head. $1,760.

Mills Novelty EXTRAORDINARY GOLDEN, 1933. The EXTRAORDINARY honored the Chicago Fair. It has massive Deco styling much like the extraordinary modern buildings on the fair grounds, and earned a "World's Fair" nickname. The initial Bell model became known as the "Gray Front." The 1934 EXTRAORDINARY GOLDEN MODEL added the GA, and a modernistic eagle on the front, giving it an "Eagle" nickname. Also made in British Penny and skill models for the U. K. $1,410.

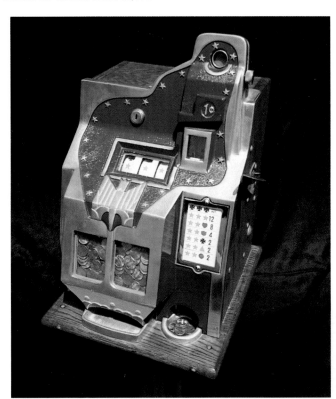

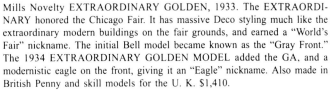

Mills Novelty Q. T., 1934. Mills entered the small Bell business, and beat the competition with the remarkable Q. T. It has a completely new and very simplified Bell mechanism. In fact, it is a wonderful collectible machine for children to play because it is easy to pull down the stamped steel handle and watch the reels spin. And make payouts if you hit. It pays just like a full size Bell. The line survived for two decades in a variety of models. This was called the "Firebird." $1,010.

Pace 1935 COMET FRONT VENDER, 1934. Pace engineering finally caught up with their Bell design, and the rotary escalator was squeezed into the front vender model by hiding a small portion of its lower area while still maintaining the integrity of the machine. Nice stunt! The additional cabinet redesign, and the detail put into both the straight payout Bell and vender models, gave this line the nickname "Fancy Front." Pace called it their "Streamlined Front Vender." $1,515.

Schwarz NR.26 MILLS VENDER, 1933. American slots became a phenomenon in Britain and Europe, with old machines shipped overseas in droves to meet the demand. C. M. Schwarz & Company of Leipzig became the leading German revamper and distributor, making original Bell machines of their own design as the Tura Automatenfabrik GmbH after 1932. MILLS VENDER used the NR.26 Schwarz coin head to accept German coinage, plus a new top. Nazi laws outlawed foreign (American!) slots in December 1933. $1,410.

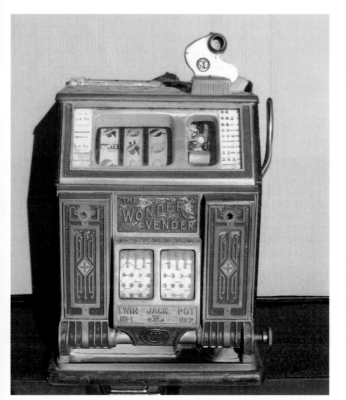

Watling WONDER VENDER, 1934. Progress at Watling. When the 3-5 payout Mills MYSTERY became so popular, Watling adopted the idea and called it their "Wonder Payout." They briefly, for one year, created a line around the machine name. This is the WONDER VENDER, nominally a Watling TWIN JACKPOT with much larger pots and the revised payout schedule. The "Large Jackpots" became the Watling standard for future models, with the pots remaining that size until the 1960s. $1,410.

Jennings DUCHESS DOUBL-JACK VENDER, 1934. Another new model introduction for Jennings in 1933 and 1934 was the DUCHESS, utilizing the same mechanism as the LITTLE DUKE turned on its side so it spun three reels instead of the circular discs. It was also available with the selective coin head as well as in this gooseneck model. This is the "Bullseye" version, named after the anti-slugging coin viewing window. $1,010.

Bull Durham TRIPLE JACK GOLD AWARD, 1934. Wide open gambling in Orange county, offshore casino boats and clandestine operations in southern California created a cottage industry in jackpot front revamps in the area. Bill Durham (his friends called him "Bull") and his Bull Durham Novelties in Los Angeles got a lot of this work, and developed unique fronts for his customers. This marvelous concoction has three pots and a Gold Award. Machine is a Jennings. $2,815.

Jennings IMPROVED CENTURY TRIPL-JACK, 1934. When the Chicago World's Fair came back for a repeat performance in 1934, Jennings came back with a new CENTURY model. Your eyes aren't deceiving you. The old operator put two left hand side vending panels of the IMPROVED CENTURY on an original CENTURY machine years ago (which means another one has two right hand panels). This is the gooseneck model without an escalator. $1,510.

Fey TWIN JAK-POT, 1934. Charlie Fey was still in slot machines in San Francisco in the '30s, now repping Pace machines and accessories, and making his own revamps and jackpot fronts. This one carries the notation "Chas. Fey Mfg. Co. S.F." in the small circle at upper left on the front. The front is the same as the replacement Pace STAR Twin Jak-Pot Front made in Chicago at the same time. Fey machines and components are highly regarded. Machine is a Jennings. $1,310.

 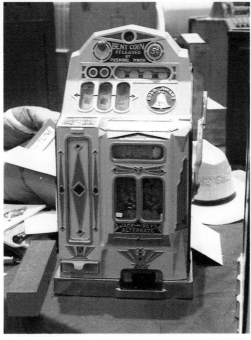 

Superior Confection GOOSENECK GOLD AWARD, 1934. The ultimate revamper was the Superior Confection Company in Columbus, Ohio. They took Mills mechanisms, intermixed with Watling parts, and cobbled up their own machines in modern looking cabinets. The GOOSENECK GOLDEN model shown here has a panel over a vending column. Take it off and fill the column and it becomes the GOOSENECK BALL GUM VENDER GOLD AWARD. Collectors love these oddball machines. $1,610.

Superior Confection ESCALATOR GOLDEN VENDER, 1934. Somehow Superior Confection was able to jam in their own patented escalator mechanism into the narrow cabinet created for the lash up Mills and Watling parts used to create their machines. The ESCALATOR GOLDEN VENDER has a Gold Award window over the Watling-style twin jackpots, and a vending column to the left that filled with roll mints. This was their Style 14. $1,760.

Superior Confection ESCALATOR VENDER, 1934. Take the same machine and eliminate the Gold Award and you have the ESCALATOR VENDER, offering roll mints in the 5¢ play model and gumballs in the penny play models. The styling of these cabinets is stunning and colorful. The same front castings are used, with the Gold Award window covered with a panel. Escalator shows the last four coins played. $1,610.

Jennings CENTURY CLUB TRIPL-JACK, 1935. A new year, and a new model of the Jennings CENTURY for private club placement, particular for WWI veterans' organizations (with the vets in their 30s and early 40s by then). The top casting of the IMPROVED CENTURY was retained, with a starburst cabinet front added. Nickname is "Century Escalator." This was the last CENTURY, as Jennings was just bringing out their famous CHIEF line, largely based on this machine. $1,410.

Watling ROL-A-TOR, 1935. Another famous machine, and one that survived in much the same styling until the 1960s. This is the original "Coin Front" ROL-A-TOR, to be pronounced ROLL-ATOR. Only the market said RO-LATER, so in May 1935 Watling changed the name to ROL-A-TOP which could hardly be mistaken. That makes the ROL-A-TOR a short population machine, and slightly preferred over later ROL-A-TOP models, although they are all highly desirable, and costly. $2,615.

45

Mills Novelty NEW Q. T., 1935. Cabinet upgrading was the vogue in 1935. The Mills Q. T. dropped its "Firebird" front for the more classic styling of the NEW Q. T. model. The "Green Front" came out in 1935, with the similar "Blue Front" arriving in 1936, followed by the CONVERTIBLE Q. T. in 1938. All have similar cabinets. Because of its styling, and the military mentality of the middle to late '30s, it picked up the nickname "Chevrons." $1,210.

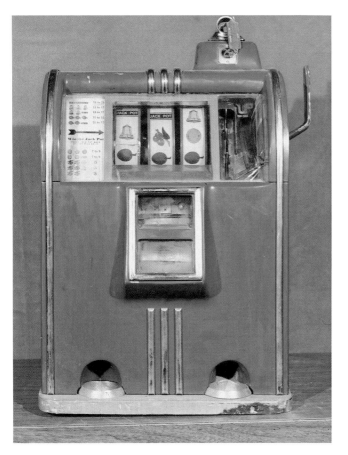

Watling ROL-A-TOP, 1935. The Watling ROL-A-TOP after the name change. They just knocked the descender off the "R" in the match plates, and continued production with the resulting "P". Later model match plates cleaned up the rough edges. Early models, such as this one, showed U. S. coins in the casting. The government said no, so the castings were changed to allegorical Roman coins. The U. S. coins model is slightly preferred over the Roman. $2,510.

Caille Bros. DOUGH BOY, 1935. Military names found favor in the mid-'30s, and the new owners of Caille Bros. (who also made Johnson outboard motors, picking up the Caille outboard in the deal) used them a lot. Their first new model machine was the DICTATOR of 1934, followed by the DOUGH BOY in 1935, named after the Americans in WWI (strangely, no one ever picked up the WWII nickname GI, or even DOGFACE, for a later slot machine). Old mechanism, new cabinet. $860.

Superior Confection RACES, 1935. The quintessential Superior Confection revamp was RACES, an enormous horse race game with small horses running in circles for a three symbol show. It was built on the frame of a Mills F. O. K. (barely visible as 2/3rds of the game at the right) plus an added vending column and the Superior Confection coin head. A truly unique and valuable machine, of hernia proportions. The DOG RACES version with two dogs and a hare has never been found. $5,020.

Superior Confection SPECIAL MYSTERY, 1935. Advanced styling was picked up by Superior Confection in Columbus, Ohio. This is a repackaging of their F. O. K. models originally made by Mills and Watling to add a twin jackpot, with the patented Superior Confection escalator head stuck awkwardly on the top, flushed to the right. The exciting play feature of this machine is that it paid on eight possible jackpot combinations. Rare. $2,410.

Jennings GOLF BALL VENDER, 1935. Adaptation of the CENTURY with a new false front that held golf balls, vending them in the manner of a mint vender. Award card on the top told you how many you got for a winner. Nickname was "Sportsman." These ran at golf courses under the watchful eye of the pro. A red ball counted double, to be turned in for two balls. Vintage golf ball venders are very desirable, particularly by golfers, and going prices reflect their popularity. $4,515.

Pace PACES RACES, 1935. One of the most significant coin machines ever made. When most slot machines sold for $100 or less, PACES RACES came out at $500 and promised to earn its cost back in 60 days. It did, often in less time. It led to a new form of payout slot called the console; larger floor machines, often with electrical features. Initial models introduced in 1934, the classic "Black Cabinet" shown here in 1935. Make bets on seven horses, and they run a race. $7,520.

A. C. Novelty MULTI-BELL, 1936. Another new starter. After Adolph Caille sold out his The Caille Bros. Co. he couldn't use his name in the business. So he started A. C. (for Adolph Caille) Novelty Company in Detroit and made a new machine. The 7-way coin head is based on PACES RACES. The play and payouts are complicated, a harbinger of the complicated payout machines of the present era. $2,810.

Seeburg GRAND CHAMPION, 1935. Horse racing was a major theme in 1935. GRAND CHAMPION was made by the J. P. Seeburg Corporation in Chicago, the coin-op player piano and jukebox makers, in an attempt to get involved in the payout slot business. It was a short lived effort. GRAND CHAMPION plays like an old 6-way floor machine. Drop the coin, pick the winning horse on the front knob, and pull the handle. If your horse stops on the line, it pays out from 2 to 20 coins. $2,810.

A. C. Novelty MULTI-BELL VENDER, 1936. Side vender model of the MULTI-BELL vended mints. The internal mechanism differed in having a check divider that diverted all the coins played directly to the cash box while it put trade tokens into the payout tube. So if you played and won you got checks to exchange in trade, and the operator and location got the cash. Rare machine. $3,420.

Exhibit Supply CHUCK-A-LETTE, 1936. Consoles took three primary forms; race running, reel spinning and electrical flashing, the latter often a lighted spinner under a top glass. The Exhibit Supply Company of Chicago elected to spin reels, with electrical backlighting of features on the top. Dice symbols on the reels at left. The 1936 model was crude; 1937 well finished with a new top on the same machine for a horse race reel version called JOCKEY CLUB. $1,010.

Groetchen COLUMBIA, 1936. New machine. From the ground up. The Groetchen Tool & Manufacturing Company (shortened to Groetchen Tool Company) was located in Chicago and made counter games. Then they opted to get into slot machines, and created (and patented!) a completely unique Bell machine with a coin carousel that returned the last coins played. So if you slugged, you got them back. Groetchen machines don't have a wide following, but some collectors adore them. $660.

Mills Novelty FUTURITY GOLDEN, 1936. The next development in Mills slots once all the basic models were available were "Feature Bells" that offered unique payout possibilities. The FUTURITY gives you ten chances to win, clicked off on the pointer dial and counter at the top. If you don't win you get all your coins back. What usually happened is that a 3 or 5 payout is made in the ten plays. But it sure kept you playing. Came in plain, Gold Award and vender models. $3,010.

Pace COMET GOLF BALL VENDER, 1936. A fast look suggests the Jennings GOLF BALL VENDER based on the CENTURY. But this is a Pace copycat version on the frame of the "Fancy Front" F. O. K. front vender. Because of the relative value of golf balls, even then, these machines were all 25¢ play. It was a way for a local club golf pro to pick up some extra money. The new front was essentially a lash-on. Golf ball machines are valuable. $5,020.

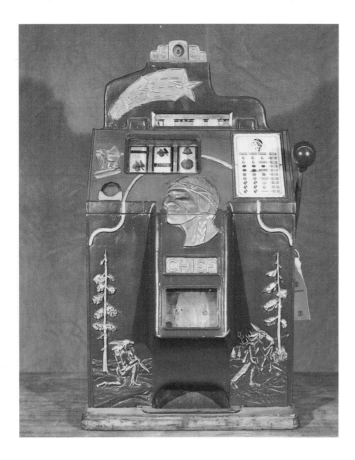

Jennings CHIEF, 1936. Introduced in 1935, the CHIEF was the firm's basic model well into the '60s. The original had a less stylish cabinet and was called the "One Star Chief" because it had a single star on its top casting. The improved coin head and design of the 1936 model gave it the nickname "Four Star Chief" based on the simulation of a comet on the top casting under the coin chute. CHIEF machines had the biggest jackpots in the business. $1,210.

Mills Novelty CHERRY BELL, 1937. The CHERRY BELL gimmick is that the cherry symbol paid more than it did on any other Mills machine, or anyone else's for that matter. A two cherry payout was 3, but two cherries and a bell or lemon paid ten. 10! The payout schedule was 3-10-10-10-14-18-jackpot. Once the CHERRY BELL was well established the payout schedule was cut back to the normal mystery payout of 3-5-5-10-14-18-jackpot. Nickname, logically, is "Bursting Cherry." $1,110.

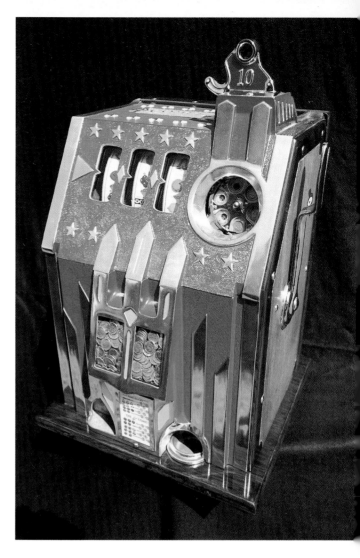

Pace ALL STAR COMET, 1936. The new look for Pace machines. After the "Fancy Front" the ALL STAR COMET presented dignified good looks. Typical of operators, they gave it a new name right away: "Plain Front." Pace serial numbers coded play features. This is serial FB-29,097-M, with "FB" meaning straight gambling Bell play, with the "M" signifying the 3-5 mystery payout format. This was also a very popular export machine. $1,210.

Buckley BONES, 1936. If 1935 was horses, 1936 was the year of the dicer. Payout dice machines were introduced by Western Equipment, Rock-Ola, Mills Novelty, Bally and Buckley. The Buckley BONES, coming out later in the year, was almost a duplicate of the Bally RELIANCE, with Buckley the original producer. Payout dice machines are well liked, and continually increase in value. The rarest is Buckley's 1939 CHERRY BONES with fruit dice, only two known. $4,020.

Pace FORTUNE MINT VENDER, 1937. Just as Mills Novelty made a FUTURE PLAY accessory that fit on most machines to beat local anti-gambling laws, Pace made an add-on device that covered the ALL STAR COMET payout cup to make it a fortune teller with copy on the reels over the fruit symbols. Special award cards in the jackpot windows explain its workings. When the territory "opened up," wham! Off came the cap on the payout tray, and out went the special award cards. $1,860.

Bally BALLY BELL, 1938. A new player entered the slot machine arena in 1938 when Bally finally created a countertop slot machine in the Bell class. The RELIANCE dice game had been introduced in 1936, but the BALLY BELL was different. Very different. It is a dual game, and takes two coins; nickel/nickel and nickel/quarter models were made. It earned the nickname "Double Bell," which was used so frequently many in the industry thought that was the actual model name. $2,410.

Mills Novelty BONUS, 1937. The major "Feature Bell" from Mills for 1937 was BONUS, with new reels and an extra mechanism that spells the word B-O-N-U-S one letter at a time in windows at the top of the machine as they are made on the first reel. But...you must hit them in order. Getting a "B" and then an "N" only gives you a "B" because you have to hit the "O" next. Spelling BONUS gave an 18 coin payout. Nickname is "Horse Head." An improved model came out in 1939. $2,410.

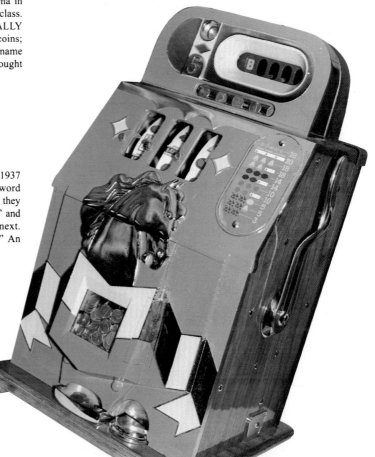

Pace KITTY, 1937. Picking up on the Mills FUTURITY and then some, the Pace KITTY is a feature ALL STAR COMET with a counter window to the right of the jackpots. There are no lemons on the first reel. The head of a black cat takes their place. Every time a "Kitty" symbol is hit a coin is dropped into a special jackpot, which holds up to 40 coins. Lose ten times in a row, counted by the window, and the pot dumps. There are straight Bell and side vender versions. Left: $2,750. Right: $3,020.

Pace TWIN ROYAL COMET, 1938. A terrific looking member of the Pace line. The ROYAL COMET floor machine, a Bell in a standing cabinet, came out in 1937. Its twin came out in 1938, and by twin we mean a doubled machine. Two ROYAL BELL slots were combined in one cabinet, sharing a center pull play handle. Play the 5¢ at left, 25¢ at right, or both the nickel and quarter machines, and the handle will spin only the machines that received coins, one or both. Exciting play. $5,520.

Groetchen COLUMBIA STANDARD, 1938. The Groetchen COLUMBIA split off into a wide range of models with fruit and cigarette reels. COLUMBIA STANDARD became the basic slot machine, the cabinet characterized by two wide stripes between the reels and play handle. The payout carousel is located in the thick trunk of the coin chute structure. Because the carousel slots weren't keyed to any specific coin size, all you needed to change coinage was a new denominator and inside mask. $660.

Watling DIAMOND BELL, 1937. One of the rare variations of the ROL-A-TOP front, the DIAMOND BELL had the same hyped 3-10 cherry payout as the Mills CHERRY BELL, plus the extra diamond cut jackpot. Both ideas were swiped from Mills. But not that cherry front casting. That's pure Watling, and absolutely stunning. DIAMOND BELL also came in a standard 3-5 mystery payout. Logically, the nickname is "Cherry Front." $3,020.

52

Groetchen GOLD AWARD COLUMBIA, 1938. Development of the Groetchen COLUMBIA led to most of the basic models of other Bell machine makers, including the Gold Award. The COLUMBIA STANDARD cabinet was modified to eliminate the twin jackpots, add another half stripe and a GA window. Payouts and the GA token went into a covered cup. COLUMBIA machines are at entry level pricing, and are a wonderful vintage slot to start a collection. They won't always be in that class. $710.

Jennings SPECIAL CHIEF, 1938. The Jennings CHIEF line followed the same prognosis as the Mills line, covering all the basic slots and then adding "Feature Bells." The SPECIAL CHIEF category covered a lot of extra models, in this case a CHIEF VENDER SKILL. The skill buttons are the three small knobs beneath the first reel, each one stopping a reel when pushed. It was the American interpretation of a predominantly British idea. $2,010.

Mills Novelty VEST POCKET, 1938. Nothing in the way of a small Bell ever beat the Mills VEST POCKET for size and performance. Inside this small cube were the complete workings of a miniaturized mystery payout Bell machine, including the payout tube and slides. A terrific tavern bartop slot. A reel cover flips over to make it look like a radio. Changes in colors, and a CHROME VEST POCKET model, kept this machine before the public well into the '50s. Nickname is "VP." $310.

Baker BAKER'S PACERS, 1939. When former Pace employee Harold Baker left the firm in 1938, he opened up his own operation in Chicago as the Baker Novelty & Manufacturing Company. What he did was revamp and upgrade Pace machines, including PACES RACES, which he put into new cabinets and re-marketed as his own BAKER'S PACERS. Mechanical improvements in the machine make it run better than a standard PACES RACES, although the Pace product seems to be preferred. $5,510.

Bally HIGH HAND, 1940. Once Bally got the hang of making reel payout machines they went off in a number of directions at once. One of the most interesting was HIGH HAND, a 5-reel poker player that allowed you to hold cards by pushbutton, and spin again to fill out your hand. With replays for winning hands. Three of a kind racked up 3, while a Royal Flush counted up to 45. Typically, the location paid off the replays in cash, while the machine remained a free play. $760.

Jennings GOOD LUCK, 1939. Floor standing electrical spinner consoles came in a wide variety of models. Most often a large color wheel was silk screened on a flat top glass and the spinner lighted the colors from below it as it came to a click stop. Jennings tripled the effect with its GOOD LUCK electrical console, spinning three Roto-Dial Spinners behind fruit symbols. Coming to a 1-2-3 stop, it paid off like a Bell machine. DELUXE GOOD LUCK had a marble finish cabinet. $910.

Jennings FREE PLAY FAST TIME, 1940. Floor standing console 3-reel machines were made by all the Bell makers in payout and free play models for various territories. The horse reel Jennings FREE PLAY FAST TIME puts a "Free Games" counter dial in the top, replacing the miniature pinball game of the FAST TIME "Skill Play" model that theoretically required skill to get the payout. You could hardly miss the ball. Fruit symbol reels created the HARVEST MOON models. $460.

Jennings BRONZE CHIEF, 1940. Styled the same as the SILVER MOON CHIEF, the BRONZE CHIEF was the last production Bell machine made by Jennings before wartime restrictions on raw materials ended slot machine production early in 1942. "By Jennings" is in its nameplate. Originally planned to be bronze coated, limits on supplies brought it out in a beige painted front. It was also the first Jennings slot machine produced after the war, with a new nameplate for THE BRONZE CHIEF. $1,510.

Jennings SILVER MOON CHIEF, 1940. Jennings seemed to like the SILVER MOON name, as they used it on a lot of different payout machines. The SILVER MOON CHIEF was just about the ultimate in pre-WWII CHIEF design. 1941 production differed in having the text "1941 MODEL" in the top casting. A more liberal payout club model was also produced as the SILVER MOON CLUB, looking virtually the same with just the name change. $1,510.

Jennings BOBTAIL, 1940. Smaller and not much more than a boxed CHIEF slot machine, the BOBTAIL console Bell used dog and hare reel symbols. Dozens of models were produced to meet all sorts of territory requirements across the country, and many have yet to be found. The same machine with fruit symbol reels and Bell payouts was called the SILVER MOON CONSOLE. Economically priced console Bells are not regarded as prime collectibles, but their popularity is growing. $360.

Pace DELUXE CONSOLE, 1940. The DELUXE COMET with its modern aircraft styled cabinet (particularly its DELUXE CHROME COMET model) with wings below the jackpots became the standard appearance of the line in 1939. A slug rejector model was added in 1940 as the DELUXE CONSOLE (strange name for a counter machine!) that picked up the styling and added the big head of the anti-slugging device. Serial FCD-51,053-M meaning Bell Comet Detector (FCD) Mystery payout (M). $1,810.

Pace COMET SLUG REJECTOR BELL, 1942. When Pearl Harbor and America's entry into WWII ended coin machine production and removed the aluminum needed for slot machine fronts, Pace packaged a small number of standing COMET CONSOLE mechanisms and components in built-up walnut stained wooden cabinets that replicated the shape and trim of the pre-war all metal originals. The wartime wooden cabinet COMET SLUG REJECTOR BELL machines are unique, and rare. $2,810.

Mills Industries CLUB ROYALE, 1945. Other pre-war Mills slots came back in postwar garb. The 1930s EXTRAORDINARY and early '40s CLUB BELL returned as the CLUB ROYALE in a Bell console floor standing cabinet. It went into private clubs and the new casinos just being built in Nevada. The eagle device of years back was dropped, replaced by an award card, and the graphic representation of a bellhop was put to the right of the reels. It gave the machine its nickname: "Page Boy." $1,510.

Mills Industries GOLDEN FALLS, 1946. The second new Mills machine after the war was GOLDEN FALLS, a handload jackpot model that found major placement in Nevada casinos built to take advantage of the end of gas rationing, with Las Vegas becoming a southern California playground. GOLDEN FALLS had the biggest jackpot of any slot machine being sold. It was only built for 2-1/2 years, but set the pattern for many revamps that upgraded older pre-war Mills slots. $1,410.

Keeney BONUS SUPER BELL, 1946. The big floor Bells came back, too. And among the biggest was a machine made in Chicago in 1941 by J. H. Keeney & Company, Inc. called the SUPER BELL. Only when the postwar version came out early in 1946 the payouts had been elevated, and the machine added BONUS to its name. The first model to hit the market was the single coin BONUS SUPER BELL with big wide reels and its own distinctive cabinet trim. 5¢ play. $610.

Mills Industries 50¢ BLACK CHERRY, 1946. War, pent up consumer demand, and the pressure on limited available postwar materials all contributed to inflation and price escalation. That included slot machines, leading to greater popularity for 50¢ and dollar machines than ever before. Casinos particularly liked them. They often required major cabinet changes. This is how Mills Industries accepted the larger coin with a gulping coin chute. Only four coins show in the escalator. $1,410.

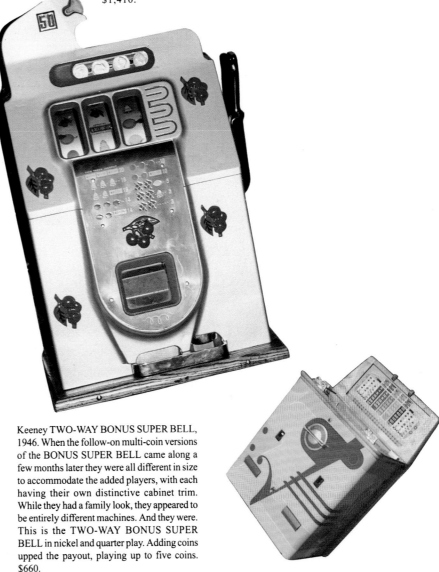

Keeney TWO-WAY BONUS SUPER BELL, 1946. When the follow-on multi-coin versions of the BONUS SUPER BELL came along a few months later they were all different in size to accommodate the added players, with each having their own distinctive cabinet trim. While they had a family look, they appeared to be entirely different machines. And they were. This is the TWO-WAY BONUS SUPER BELL in nickel and quarter play. Adding coins upped the payout, playing up to five coins. $660.

Mills Industries BLACK CHERRY, 1945. The WWII years changed the coin machine industry a lot. It even changed the Mills Novelty Company name. A novelty company could hardly make aircraft components and food service equipment. So "Novelty" became "Industries," and the name stuck after the war when Mills slots were among the first to go back into production. BLACK CHERRY was a basic pre-war "Halftop" escalator Bell with a new postwar cabinet. It was produced for three years. $1,210.

Bally TRIPLE BELL, 1946. Bally Manufacturing Company in Chicago got their toes wet in consoles just before WWII, and jumped into the pool right after the war. Their first was a model called DRAW BELL in 1946, followed by a large floor machine called the TRIPLE BELL. One Bally Bell mechanism was coupled to three coin chutes on which you could play up to three coins in each. So it was a triple triple. TRIPLE BELL is Bally Model 470. $810.

Mills Industries 50¢ GOLDEN FALLS, 1946. The 50¢ version of GOLDEN FALLS utilized the same form of outreaching coin chute as the 50¢ BLACK CHERRY. To give you an idea of how many Mills slot machines were produced over the years, this is serial 529,573, meaning over half a million Mills Bells since the LIBERTY BELL of 1907. That's why these machines are collectibles, and why finds are still being made (and will be for many years hence). $1,510.

Mills Industries NEW Q. T., 1947. The Mills Q. T. came back after the war, but only briefly. It was completely restyled to have a modernistic flowing front, utilizing much the same mechanism as the prewar model. When it came out a sales arm of Mills called Bell-O-Matic handled distribution, so the Mills names virtually disappeared from their slot machines. It was discontinued in September 1949. It picked up the military nickname "Hash Marks." $1,210.

Jennings MONTE CARLO CHALLENGER, 1948. The only successful console Bell made after WWII with a standard slot machine mechanism, the Jennings CHALLENGER came out in 1947 with a wood rim top and backglass. The 1948 versions replaced the wood with metal rims, and added models. This is the 500 coin "Super Jack Pot" MONTE CARLO version. Many more models were made for the British market into the '60s. $2,615.

Mills Industries JEWEL, 1947. Mills changed the look of slot machines when they introduced the JEWEL in 1947. It quickly garnered the nickname "Hightop," making the earlier escalator Bells "Halftops." The mechanism was largely unchanged, but the appearance alteration was dramatic. "Hightops" became the best know slot machine configuration in the world in the 1950s and '60s. The JEWEL was the first of the line, and all subsequent machines were made in the same format. $1,410.

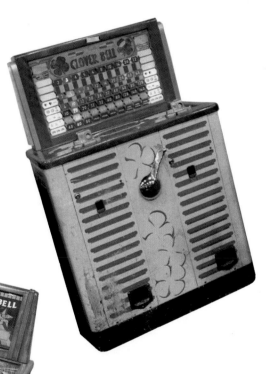

Bally DELUXE DRAW BELL, 1947. Bally brought back their DRAW BELL idea in 1947, adding an extra draw feature that allowed a player to play up to five more coins, holding the reels selected in an attempt to better the payout with five extra pulls. The Bally floor consoles carved out a whole new market for slot machines in roadhouses and taverns, challenging the laws everywhere. These big machines with lightup features are fun to play and exciting to watch. $510.

Bally CLOVER BELL, 1949. As the '40s ended the big postwar console machines got bigger, and the play more complicated. The Bally CLOVER BELL was created to compete against the Keeney TWO-WAY BONUS SUPER BELL, with additional features. Added coins built up the odds, and lighted symbols pay extra awards. These large consoles have never caught on. It takes a newer generation of collectors unafraid of electricity to appreciate their many merits. They are a coming collectible. $810.

Bell-O-Matic BONUS, 1948. Mills produced the "Hightop" in just about every prewar feature Bell format they made (except the FUTURITY and FUTURE PLAY, both deemed "too conservative" for casino locations), plus a bunch of new ones. The old "Horse Head" NEW BONUS of 1939 came back as the BONUS "Hightop" in 1948. The BONUS "Hightop" remained a Bell-O-Matic (see Mills) offering well into the '50s, even after the factory moved to Nevada to comply with the Johnson Act of 1951. $1,770.

Jennings STANDARD CHIEF CONSOLE, 1947. The basic postwar STANDARD CHIEF Jennings design used the prewar "Chief" mechanism with some mechanical improvements in a new highly streamlined polished aluminum cabinet, introduced in 1946. In 1947 the STANDARD CHIEF was also available in a floor standing cabinet as STANDARD CHIEF CONSOLE, shown here. Light-up features were added to other machines in the line as CLUB CHIEF and SUPER DELUXE CLUB CHIEF. $2,820.

Bally TRIPLE DRAW BELL, 1950. The tendency to get bigger and bigger, adding features every time, soon made the Bally consoles gargantuan. The TRIPLE DRAW BELL is the logical extension of this thinking, making the comfortably sized DRAW BELL into a light-up monster. But what fun! Three can play; there are three coin heads and payout cups, all keyed to the single set of three reels, plus hold-and-draw features for each player. What a wallop! $810.

Bell-O-Matic WILD DEUCE, 1950. The last Mills "Hightop" produced in Chicago prior to moving the entire gambling machine operation to Nevada once the Johnson Act outlawed slot machines and Illinois law prevented their manufacture in the state. It is the only "Hightop" painted a bright canary yellow as a model color. The side panel says "Deuce 2 Wild" and the top panel says "Wild Deuce," so the actual name of the machine is often confused. $1,410.

Bell-O-Matic 21 BELL, 1949. The basic Mills "Hightop" by the end of the '40s was the triple-7 "21," with reel symbols having watermelons and outlined "7" symbols over some of the fruits. Hitting three 7 symbols paid a Special Award, $10 in the case of a nickel machine. By making payouts more liberal the roadhouses and casinos that ran slots attracted more players. By 1950 the 7-7-7 became the 21 STANDARD, and a gleaming CHROME 21 model was added. $1,210.

Pace SUPER JACKPOT 4TH REEL, 1950. The last Chicago Pace slot was called the "8 Star Bell" because of the language on a front panel. The actual basic model is SUPER JACKPOT. This specific version adds a 4th reel left of the first reel to increase the payout odds, making it the SUPER JACKPOT 4TH REEL, dollar play. Pace tried all sorts of mechanical stunts to increase play, even offering some models in transparent cabinets like this so the players could follow the action. $2,015.

Keeney PYRAMID, 1950. In 1949, in order to get a new model, Keeney turned the BONUS SUPER BELL mechanism on its side, facing the show forward, to create the PYRAMID console Bell, a true heavyweight. When the Johnson Act was imminent it was converted to casino criss-cross play with many bar winners. Orders went to Nevada, Maryland and a number of the casinos that were illegally operating in Hot Springs, Arkansas. Nickname is "Egyptian," for the figure on the cabinet. $1,810.

Keeney CRISS CROSS, 1952. Keeney kept making slot machines in Chicago, and ultimately merged with Jennings, with all production remaining at Lake Street on the west side. The CRISS CROSS was a stopgap machine, getting orders only as long as Mills and Jennings Bell machines were in short supply. It came in both console, on a stand, and tabletop models, although the latter is a misnomer as these are very large and heavy machines, particularly as collectibles. $1,410.

Buckley KENTUCKY DERBY, 1950. The anti-gambling laws against slot machines didn't cover non-payouts. All sorts of payout machines were converted to "free play" models, in which the "wins" were tallied as replays on a counter, to be played off or, clandestinely, paid off in cash against the numbers shown. Buckley Manufacturing Company of Chicago converted their famous TRACK ODDS electrical console, making it a replay "For Amusement Only" machine as the KENTUCKY DERBY. $460.

Jennings DOUBLE STAR CHIEF, 1950. Casino locations wanted something unique and rewarding to players willing to take exceptional chances. The DOUBLE STAR CHIEF more than doubles the payouts on 6 reels by coupling two machines together to be operated by one handle pull. Double machines tend to be rare and command high prices as they are two machines in one, plus the uniqueness of the coupling and payouts. A pair of Chief busts can be seen over the two jackpots. $3,420.

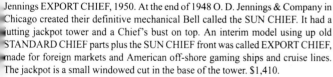

Jennings EXPORT CHIEF, 1950. At the end of 1948 O. D. Jennings & Company in Chicago created their definitive mechanical Bell called the SUN CHIEF. It had a jutting jackpot tower and a Chief's bust on top. An interim model using up old STANDARD CHIEF parts plus the SUN CHIEF front was called EXPORT CHIEF, made for foreign markets and American off-shore gaming ships and cruise lines. The jackpot is a small windowed cut in the base of the tower. $1,410.

Jennings ENLARGED JACKPOT SUN CHIEF TIC-TAC-TOE, 1950. The liberal Tic-Tac-Toe payout pioneered by Buckley in Chicago was soon picked up by all of the manufacturers. It became a standard payout format in the Jennings line. By the end of 1950 Jennings also enlarged the jackpot on the SUN CHIEF, with the new pot window taking two segments of the tower where it only took one in the past. Tic-Tac-Toe payout symbols are in the pot window. $1,610.

Jennings BUCKAROO, 1954. Jennings continued making machines for Nevada in Chicago, and developed the 4-reel BUCKAROO for The Nevada Club at the end of 1954. Adding another reel enabled the machine to pay a 5,000:1 jackpot. A very desirable and valued collectible slot as it has light-up features, and offers the excitement of a high payout rate. Jennings produced the SUN CHIEF models and their advanced variants until the middle 1960s. $2,810.

Jennings BRONZE CHIEF JOKER, 1953. The JOKER approach of getting around the Johnson Act of January 2, 1951, outlawing slot play except in states where it was legal, was developed by P & M Enterprises in Lander, Wyoming. You put money in the machine, in this case in the base, and ran up replay points in a small window to the right. Then pulled the handle until they were played off. If you came out ahead, you got paid over-the-counter by points. Not very desirable collectibles. $1,010.

Games HUNTER, 1955. The anti-gambling machine laws led to electronic uprights. Essentially electrical stepper machines based on odds building Bingo-type pinball games, they score by flashing lights behind symbols on a backglass, earning replay scores. Leading makers were Keeney, Auto-Bell and Games Incorporated, all in Chicago. HUNTER was among the first. Because they don't pay out, and are very electrical, these games are shunned. Yet they are quite exciting. $360.

Buckley CRISS CROSS ELECTRONIC POINTMAKER, 1956. Another attempt at beating the law, this in the form of a light flashing 3-reel machine. Buckley Manufacturing Company took their mechanical CRISS CROSS Bell slot machine, hard wired it to display lights behind the symbols in the manner of a reel spin, and rack up points as the POINTMAKER. Later versions had this sleek cabinet. Some day these will be regarded as pioneer electronics, and valuable. But they aren't now. $360.

Keeney WILD ARROW, 1960. The greatest advances in electronic uprights were made by Keeney in Chicago, who produced them into the '70s. Of limited popularity as replay machines in the U. S. (because actual payouts were illegal) they were highly popular in Great Britain in their payout versions, as shown. Keeney added the PANOSCOPE, a computer generated three window display of fruit symbols at the top that recreated the reels of the past. Adding coins multiplied payouts as collectibles. $210.

Ramsdell 6 BELLS, 1960. The laws of Maryland allowed slot gambling in four counties, but reduced the number of machines annually until they would be gone. So revampers created machines with multiple players to keep the play count up while the machine count went down. Ramsdell Engineering Corporation of Baltimore, Maryland, converted prewar Mills FOUR BELLS consoles into 6 BELLS with upgraded components, added payout mechanisms, and new top glass. Just catching on. $1,010.

Bally MONEY HONEY, 1964. The landmark slot that changed the industry, and led to the modern gaming machine. The electromechanical Bally MONEY HONEY with its hopper payout permitted all sorts of features, multipliers and large payouts in a single countertop machine, doing everything the massive consoles did only a few years before. Its styling revolutionized slot machine design throughout the world to this day. Ballys are highly collectible where allowed by law. $1,410.

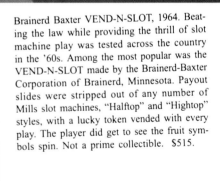

Jennings CLUB CHIEF TIC-TAC-TOE, 1960. The British market was a windfall to the struggling American slot machine industry, as the Brits bought in droves, requiring new models and revamped payouts to fit the new coinage. Existing models with new names, such as GOVERNOR, GRUBSTAKE, LUCKY PUNTER and others, including this CLUB CHIEF TIC-TAC-TOE in 6 pence coinage, served the Brits for a dozen years. They are now coming back to the U. S. in the same droves. $1,510.

Brainerd Baxter VEND-N-SLOT, 1964. Beating the law while providing the thrill of slot machine play was tested across the country in the '60s. Among the most popular was the VEND-N-SLOT made by the Brainerd-Baxter Corporation of Brainerd, Minnesota. Payout slides were stripped out of any number of Mills slot machines, "Halftop" and "Hightop" styles, with a lucky token vended with every play. The player did get to see the fruit symbols spin. Not a prime collectible. $515.

Jennings CONTINENTAL BUCKAROO, 1964. When the Bally machines first showed up they took a while to penetrate the market. But the writing was on the wall. Everything else had to be modernized. The Jennings SUN CHIEF and ELDORADO CHIEF lines came in for their share with the front opening CONTINENTAL, made in a variety of models. The Nevada Club ran CONTINENTAL BUCKAROO machines. Similar ELDORADO large marquee models were assembled in the Bahamas for British use. $2,810.

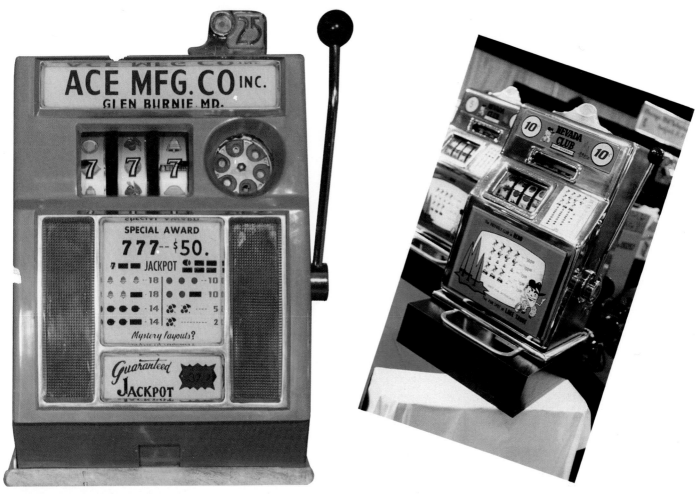

Ace SPECIAL AWARD, 1964. All of the mechanical slot machine makers scrambled frantically to counter the Bally threat. Ace Manufacturing Company, Inc., in Glen Burnie, Maryland, produced front opening machines based on models created in Nevada in the late '50s and early '60s before the Ace organization had moved back to Chicago, with a Maryland spinoff. They found a ready market in the local counties, but sputtered elsewhere. Machines are usually found in Maryland. $860.

Baldecchi NEVADA CLUB 4 REEL, 1967. With its enormous inventory of older Jennings slots, The Nevada Club sought an economical upgrading solution. They commissioned the Baldecchi Manufacturing Company in Reno, Nevada, to repackage their machines. The Baldecchi all steel case made the machines virtually silent, and look newer. They went in storage, and are only now coming out for sale to collectors. Great slot for a rec room. 10¢ play nickname is "Green Front." 5¢ is "Blue Front." $1,210.

Jennings GALAXY MECHANICAL JACKPOT, 1966. Jennings finally accepted the unacceptable and restyled in accordance with Bally principles, and even included a hopper payout. GALAXY became the leadership line, produced in a wide variety of models in mechanical, electromechanical and solid state versions. They are just beginning to enter the collector market. The Jennings GALAXY MECHANICAL JACKPOT still had a mechanically dropped jackpot, a growing rarity by the middle '60s. $1,110.

Mills Bell-O-Matic MODEL M, 1968. Mills finally caught up with current styling at the end of the '60s, with a disastrous machine called the MODEL M. It works well, looks good, and is inexpensive as a collectible. It just never caught on in the casinos. The 1968 Mills MODEL M "M-Head" was the firm's first multiplier, made in a variety of models from progressive to multiple, double and single coin versions. A conventional and playable machine, and often a collectible buy. $860.

TJM 400 5-LINE CRISS CROSS, 1979. Jennings went through a series of owners in the '70s and the '80s, with all having a hand in the 400 and later 700 SERIES. Jennings And Company in Chicago became a part of TJM Corporation in Elgin and elsewhere in Illinois, later OTX Corporation in Nevada and finally the existing Mills Jennings Company in Nevada. This machine carries custom glass for Caesar's in Atlantic City. These are reliable as rec room game machines. $1,110.

Jukeboxes

The jukebox has been one of the most enduring fads in vintage coin machine collecting, and is probably the second most desirable collectible coin-op, both in popularity and value. Long before people around the country (and, following that, the world) got interested in old slot machines and counter games, soon discovering vending and scales and arcade and whatnot, people of means and parents of teenagers saw the virtue of picking up an old jukebox and keeping it in the basement rec room, out at the cottage or somewhere that kept the kids at home and good old records playing. Sturdy and durable (they had to be because they were made for public places), strong sounding and often terrific looking (particularly with bubble tubes, and changing patterns of colored light) these pieces of commercial come-on turned into great items of furniture. The boy or girl who could invite other kids over to drop nickels into a jukebox and dance (or just listen!) was a fortunate child indeed. The whole phenomena was a set-up of preservation for the vintage coin-op collectors of a later day, for it was these home saved machines that initiated the jukebox collecting trend.

Jukebox collecting has a history all its own. Jukeboxes, first called automatic phonographs, entered the marketplace around 1928 with the availability of amplified sound, a side benefit of radio. Coin operated phonographs were made before then, with the Edison machines joined by Mills Novelty, Rosenfield Manufacturing, Regina and Caille Bros. Co. machines made around the turn of the century and into the teens and '20s. But Prohibition just about killed that business. With no locations serving beer and drinks, dancing went into ballrooms rather than bars with live bands providing the music. But in spite of the sparsity of places to play, the amplification of music was such a novelty that the coin-op music machines reentered the marketplace even though there were few places to go. Sales struggled for 5 or 6 years. Then came the November 1932 election of Franklin Delano Roosevelt and a pledge to repeal Prohibition. It happened in May 1933 with beer, with whiskey coming under the new laws by the end of the year. By late 1934, the automatic phonograph was the best selling coin machine being made. Public pay-to-play music came back as thousands of taverns and pouring spots were established all over the country. Happy days were here again!

Established machine makers got into the game, including Wurlitzer, Seeburg, Rock-Ola, Mills Novelty (a major jukebox pioneer, and an early amplifier), Gabel, and a host of others, to create the classic machines of the 1930s. Development was staggeringly rapid, with the number of records handled quickly going from 10 to 12 to 20, 30 and beyond with automatic systems, wall boxes and speaker systems proliferating. Within five years of their entry, jukeboxes had taken command of a major portion of the coin machine industry. It was about this time that the Big Bands—which had an enormous public following—became the backbone of the automatic phonograph business, and vice versa with the new machines—now starting to be called jukeboxes around 1940 (an African-American term the industry hated, and would not accept for years)—supporting the bands. The term "Juke Box" traces it's origins to the Southern Black community where road houses were called "Jook Joints." "Colored" music was played there by both black and white musicians. It was the musicans themselves who applied the term to the machine, the "Juke Box," that played their records in these Southern establishments. Despite the dispute over the name, for the jukebox and the musicans the partnership was an evergreen two way street.

After World War II the machines came back even stronger as the men and women in service returned home to go out a lot, with the most highly collectible machines being the 78 RPM models produced in the 1945-1955 period. These were the machines that made it to the suburban rec rooms and, finally, into the hands of a dedicated and enthusiastic army of collectors. In the beginning most private owners had one box which

Edison MODEL M, 1896. Thomas Alva Edison, the inventor of the cylinder phonograph (and electric light, the movies, and wads of other things) saw the phonograph as a business tool. He was late to see it as an entertainment device, but finally caught on. Edison Phonograph Works of Orange, New Jersey, finally made the AUTOMATIC PHONOGRAPH in 1891. The curved glass MODEL M followed in 1896. The internal flexible tube connects to earphones or a horn speaker. $15,050.

they used and abused, and that was enough. But little by little others saw the beauty and charm of the machine, and started to pick up more boxes, and ultimately speakers and remote wall and countertop selectors.

Which is one of the more interesting facets of jukebox collecting. Some collectors don't even have a jukebox, preferring to specialize in speakers, or bar boxes. Classics of their genre have become recognized, with the Seeburg 3W-1 (3W for "3-wire system") wallbox of the 1950s, with its flipping tables of record selections, becoming a '50s pop culture ikon of sorts. You can see it's shape recreated to this day in novelty radios and telephone directories. Fortunately, the original location population of these wallboxes, and others of the period, were so high you can still pick them up in antique shops and malls at reasonable prices. Other true classics include the Wurlitzer "Strike Up The Band" Model 250/350 speakers of 1939, the classic Seeburg "Teardrop" speakers of the late '30s and immediate post-WWII period, and the legendary (because of its flashy beauty and the fact it was produced in such short supply) Wurlitzer 580/580A wall mount lighted speaker and selector combination of 1942. Some people even specialize further by collecting jukebox manuals, wiring sets and other paraphernalia.

the collectibles of the day, starting in the late 1960s and into the 1970s, jukeboxes began to slowly crawl out of the woodwork and rec rooms. One of the reasons for their slow start as collectibles was the fact that they were still in operation in locations all over the country, and the supply seemed endless. In the middle 1970s you could still go to an operator in most major or even smaller metropolitan areas, and pick up their old machines for $100 each or less when they moved new equipment into the routes and music locations. Stories of warehouse finds in the '60s and early '70s told of stashes of jukeboxes for a few bucks each, with some pickers even leaving the jukeboxes behind to get at the slots and arcade games. The coming of 45 RPM records in the '50s had made the older 78 RPM machines all but passe, enduring in those home based rec rooms where the records were still in ample supply and the kids loved to dance to Golden Oldies. Jukeboxes just weren't regarded as antiques as slot machines and trade stimulators were. They just sort of hung around as old outdated machines.

Then something happened. By the late '70s, when 45 RPM jukes were still being picked up for a couple of hundred dollars or so, the old 78s began to rise in value. Suddenly the machines that played them were collectibles, and antiques. When collec-

Columbia Phonograph TYPE AS GRAPHOPHONE, 1897. First coin-op phonograph that really brought music to the people. TYPE AS was found in saloons, ice cream parlors, amusement centers and the many other places that embraced the new technology of recorded music. Marquee carries the name of the Edison cylinder selection. TYPE A was made in Washington DC in 1896. When Columbia Phonograph Company moved to New York City in 1897 the TYPE AS decal showed the new location. $5,415

Edison ACME, 1906. Automatic phonographs were big business for Edison, in addition to office transcription equipment. With the rise of AC current usage (which Edison fought like a tiger because it was promoted by Westinghouse) in both major metropolitan and smaller cities and towns, the Edison ECLIPSE DC current and the Edison WINDSOR battery operated automatic cylinder phonograph models led to the AC current ACME model. They all had the same cabinets. $12,050.

But for all that, it's the boxes themselves that come in for the greatest attention, for they give a coin-op machine collector something that no other vintage machine can provide: selectable music! Jukeboxes are music machines first, and coin machines second. That alone is probably why they were on a collecting track all their own, and seemed to remain quietly underground for so many years. But once vintage coin-ops became

tors discovered how few of the 1941 Wurlitzer Model 850 "Peacock" jukeboxes remained, they rose in value from $200 to $500, jumping to $1,000 and then screaming forward in the 1980s past the $10,000 mark. One recently sold at auction for $32,000. The immediate post-WWII Wurlitzer 1015 playing 78 RPM records became the favored collectible for its classic curved cathedral top, and values raced upward. By the early 1990s the machine had broken the $10,000 mark for a prime condition collectible, where only a dozen years before they were junkyard fodder. This was particularly unexpected because the Wurlitzer 1015 remains one of the highest population jukeboxes ever built, with no semblance of rarity to it. It's the look, and desirability, that counts, in one of the strange paradoxes of vintage coin machine collecting. Even today, with half a dozen different replications of the 1015 being reproduced, including both 45 and CD versions made by the surviving Wurlitzer plant in Germany, and with over 10,000 of the newer versions already sold, the machine remains a viable acquisition. In the meantime, the original 1015 remains a classic, and holds its value.

These are the exceptions. The rule is that jukeboxes are worth about ten times today what they were a little over a decade ago. Does that mean that all jukeboxes are valuable? In the sense that they are collectibles, yes it does. But as far as the dollars are concerned this is not necessarily true. Once you get past their public armored exteriors and cushioning against shock (such as slapping, or kicking) jukeboxes are delicate machines. To be acceptable as collectibles they have to deliver their advantage and provide music at command. Jukeboxes are golden, meaning they all have some value. But they are also an item that cries out for restoration. Junkers with severely damaged or missing parts have little value as it takes a lot to bring them back into shape. But if you have a machine that appears to be in good restorable condition, with replaceable parts missing, there is a whole underground of parts dealers and restorers that cater specifically to the jukebox market. Conversely, classic and desirable jukeboxes command high prices. For instance, a Seeburg 100 SELECT-O-MATIC 45 RPM that sold at dealer retail for $450 in 1983 sells for anywhere from $1,200 to $3,200 today depending on the model and condition.

There are signs that this collecting fad has levelled off, with prices no longer on the climb. But they have held their level and stayed there for almost a decade now, so there is no apparent major drop off. In short, if you want some great vintage music around the house, consider getting a vintage jukebox for 78s and 45s, or even a repro or one of the better replications, with the latter available also in CD. And if you've got an old one you want to sell, learn what you have and get an idea of rarity, desirability and value, and then enter the marketplace with confidence. Or if you are attracted to the machine and want to keep it around the house, consider a restoration to get the best out of it. There is nothing quite like the sound and the look of a vintage jukebox blaring out one of your favorite nostalgic tunes.

Jukeboxes are still being produced in volume, and it is likely that the CD players of today will be the collectibles of tomorrow. The story of jukebox collectibility is just beginning.

Mills Novelty FORTUNE TELLER PHONOGRAPH, 1907. Originally confined to phonograph makers, coin machine makers jumped into the new machine class in the early 1900s. Mills Novelty Company in Chicago developed patent positions with both cylinder and disc machines. The FORTUNE TELLER PHONOGRAPH was a modified model of the Mills NEW AUTOMATIC PHONOGRAPH (CYLINDER MODEL) that spun a dial in the inside left that revealed a fortune every time the recording was played. $7,515.

Columbia Phonograph TYPE BS EAGLE GRAPHOPHONE, 1898. Smaller, lighter, cheaper, and the most popular of the Columbia coin-ops. It sold new for $10. It played a 2-minute cylinder for a nickel, big money in those days. The marquees were changed when the recordings were. This one says: "Drop a nickel in the slot and hear BELLE OF CHICAGO." Most desirable, and rare, is a highly nickel-plated model that originally sold for $13. $2,810.

Regina HEXAPHONE STYLE 101, 1909. An established music box maker, with experience in coin-op music boxes and gum venders, The Regina Company of Rahway, New Jersey, saw opportunities in automatic phonographs and created their HEXAPHONE. STYLE 101 hid the speaker behind a grille, while STYLE 102 (1911-1915), STYLE 103 (1914-1921) and STYLE 104 (1915-1921) have an elegant wooden horn built in to the cabinet. These are the last of the cylinder machines. Fun to watch. $7,020.

RCA MODEL CE-29 AUTOMATIC VICTROLA, 1932. With many amplified sound patents to their name, the Radio Corporation of America saw the commercial coin-operated automatic phonograph as a potential cash cow. So they made one. It was a disaster. While the CE-29 was a good machine, RCA knew nothing about coin machine distribution. They even introduced their coin-op at the radio trade convention in May, not the January coin machine show. It died aborning. Non-selective. Rare. $1,610.

Gabel AUTOMATIC ENTERTAINER, 1929. Electricity was no mystery to Gabel, with the large GABEL'S AUTOMATIC ENTERTAINER models of the early '20s having powered turntables. So when electrical amplification came along, Gabel grabbed it and added it to his now standard 78 rpm system. The AUTOMATIC ENTERTAINER of 1929 looked much like the earlier acoustic models, but had pioneer amplification. It picked up the nickname of the "Amplified Entertainer." $3,810.

Gabel GABEL'S AUTOMATIC ENTERTAINER, 1917. John Gabel was a true pioneer of 78 rpm disc play, creating his first ENTERTAINER in 1905 when he was still making slot machines in Chicago as the Automatic Machine & Tool Company. He applied for additional patents on an "Automatic Talking Machine" in 1916, making the newer version of his GABEL'S AUTOMATIC ENTERTAINER in 1917 at his new Gabel's Entertainer Company. Machine has an automatic needle changer. $4,410.

Caille Bros. CAIL-O-PHONE STYLE B, 1908. Wherever Mills Novelty went, Caille Brothers Company of Detroit, Michigan, was sure to follow, and vice-versa. CAIL-O-PHONE is a cylinder machine promoted as durable enough for public places and arcade locations, with Caille suggesting others weren't. The STYLE B runs by a spring motor that is quickly wound up by an electric motor in the cabinet. STYLE A operates on AC-current as a plug-in at a higher power cost. $3,510.

Holcomb & Hoke ELECTRAMUSE SUPER-TONE, 1929. Holcomb & Hoke Manufacturing Company in Indianapolis, Indiana, a popcorn machine and coin-op bowling game maker, saw potential in automatic disc phonographs. The ELECTRAMUSE GRAND came out in 1928. It automatically changed and continuously played 10 to 12 records (depending on their thickness), but you couldn't select them. The electrically amplified ELECTRAMUSE SUPER-TONE model came out in 1929. $3,515.

Capehart MODEL NO. 3 JUNIOR, 1931. Homer E. Capehart, sales manager at Holcomb & Hoke, went off on his own to make his fortune in automatic phonographs. He did. A couple of times. His Capehart Corporation in Fort Wayne, Indiana, prospered, failed, prospered, and failed again under a variety of names. Capehart JUNIOR has an automatic changer that handles ten 10- or 12-inch records, played in sequence, non-selective. You can repeat, or throw a switch for the next record. $1,210.

Seeburg SYMPHONOLA MODERNISTIC MODEL D DELUXE, 1936. From the sublime to the almost ridiculous, the Seeburg SYMPHONOLA MODERNISTIC MODEL D DELUXE carried the "departure" design theme to its outer limits. It's nickname was simply "Modernistic." The MODEL C looked the same, having 5 tube amplification and a single speaker. The MODEL D DELUXE upped that to 7 tubes and two speakers. 12 record selectivity. $2,810.

Mills Novelty DELUXE DANCE MASTER MODEL 886, 1936. Mills also amplified in 1927, and developed a "Ferris Wheel" mechanism that spun records around to the playing arm. Over the years not much changed but the cabinets, plus added features, such as continually improved selection systems. The DANCE MASTER MODEL 876 of 1934 added a 12 selection dial. The DELUXE DANCE MASTER MODEL 886 of 1936 came with single or triple slots in a fine wood veneered cabinet. $810.

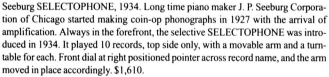

Seeburg SELECTOPHONE, 1934. Long time piano maker J. P. Seeburg Corporation of Chicago started making coin-op phonographs in 1927 with the arrival of amplification. Always in the forefront, the selective SELECTOPHONE was introduced in 1934. It played 10 records, top side only, with a movable arm and a turntable for each. Front dial at right positioned pointer across record name, and the arm moved in place accordingly. $1,610.

Wurlitzer SIMPLEX MODEL 10, 1933. Coin-op phonographs took off when Rudolph Wurlitzer Manufacturing Company of North Tonawanda, New York, a musical instrument and band organ producer, hired Homer E. Capehart as VP sales, bought out the patents of the Wilcox Ampliphone Company in Chicago, and leaped into manufacturing and sales. DEBUTANTE was the first model, duplicating the Wilcox AMPLIPHONE. MODEL 10 was a test pre-production 10 record selective SIMPLEX run. $2,215.

Rock-Ola NIGHT CLUB, 1936. Interested in music, David C. Rockola of the Rock-Ola Manufacturing Corporation in Chicago entered the automatic phonographic field in 1935 by acquiring numerous past and recent amplified talking machine, automatic changer and selector patents. The 12 selection NIGHT CLUB of 1936 is an advanced model of the original Rock-Ola MULTI-SELECTOR with greater amplification, lower price than its companion REGULAR. 5/10/25¢ slides. $1,610.

Seeburg 1936 HIGH FIDELITY SELECTOPHONE, 1936. The Seeburg SELECTOPHONE underwent a series of technical upgrades and cabinet redesigns to keep up with the rapidly changing industry. The 1936 HIGH FIDELITY SELECTOPHONE was available in three color combinations, with the Mandarin Red with Gold and Black trim shown. Other basic colors were Oriental Matched Walnut and Nile Green. 10 selections by dial at right. $1,810.

Seeburg SYMPHONOLA MODEL B, 1936. The 12 selection Seeburg SYMPHONOLA is a larger and more complex machine than the SELECTOPHONE, with its selection settings dial set by number on a highly visible horizontal panel so you are no longer looking inside of the machine. The incredible Deco styled cabinet was called a "departure" design with striped "fins," foretelling the automotive grillwork designs of the 1950s. $2,810.

Wurlitzer 716 SIMPLEX, 1937. Under the highly promotional guiding hand of Homer Capehart, Wurlitzer literally pushed itself into the dominant position as America's premier automatic phonograph maker. Breaking with the tradition of the stained "woodie" Wurlitzer SIMPLEX cabinets of the past, the 716 sported a multicolor painted one. It was the last of the full size all wood boxes. 716 SIMPLEX had a 16 selection capacity and dial; 712 SIMPLEX had 12. $2,510.

Rock-Ola MONARCH 20, 1938. Rock-Ola kept the heat up in the industry, adding colorful lighting effects in a 20 selection automatic phonograph in 1938. Only three years of development brought the coin-op phonograph from a dull stained wood box to a colorful purveyor of entertainment. The Model MON-20 brought the automatic phonograph out of the shadows and made it an integral part of a location decor. It could be a partial reason the term "Juke Box" (derived from the Southern Black slang "Jook Joints" for the road houses which operated these machines) found favor. $1,510.

AMI STREAMLINER, 1938. A starter of the plastics revolution in jukebox design, STREAMLINER features a backlighted grille, lighted plastic panels and a 20 selections "dashboard" display. After years of dull boxes from Automatic Instrument Company in Grand Rapids, Michigan, STREAMLINER was their most successful model. Model SC-99 has three coin chutes for nickel, dime and quarter. Model SC-109 shown has twenty nickel slots that selects records by their chutes. $3,410.

Cinematone PENNY PHONO, 1939. One of the best looking jukeboxes ever made, it is almost unbelievable it was produced in 1939. Styling is Moderne, a uninhibited corruption of the more rigid Art Deco. Cinematone Corporation was located in Hollywood, California. The PENNY PHONO played its own 12-inch, 20 selection (10 on each side!) variable rpm records. Model A-1 fit on a counter, with the console Model B-1 shown. Rare and valuable. The records, too. $3,615.

Gabel JUNIOR ELITE, 1936. More than any other old line automatic phonograph makers, Gabel kept up with the times. But they had limited facilities, so were never a major producer. Looking like a Wurlitzer with its 12 selection "Programatic" dial, the ELITE came in a fine wood cabinet of oriental walnut, prima-vera, zebra wood and American walnut with "ebony" (meaning black) trim. It has one of the most beautiful cabinets ever made for an automatic phonograph. $1,810.

Rock-Ola STANDARD 20, 1939. Plastic makes the box! In two years lighted plastic panels were all but mandatory to proclaim a modern automatic phonograph, and those that didn't have it were left in the dust. The Rock-Ola Model ST-20 STANDARD 20, for 20 selections, was a member of the 1939 "Luxury Light-Up" line that also included Model DE-20 DELUXE 20 with twice the lighting and a different cabinet and grille design. The Golden Age of jukeboxes was beginning. $2,210.

Jacobs Novelty 616 ILLUMINATED FRONT, 1939. New light-up jukeboxes made everything produced a year or two earlier look obsolete. A cottage industry sprang up making new cabinets, lighted plastic fronts and grilles for older machines. Jacobs Novelty Company of Stevens Point, Wisconsin, was one of dozens of revampers and kit makers. This 16 selection Wurlitzer 616 "woodie" of 1937 is all gussied up. Revamping sometimes makes it difficult to identify surviving 1930s jukeboxes. $2,210.

Mills Novelty ZEPHYR MODEL 900, 1938. Yet another redesigned cabinet for the aging Mills "Ferris Wheel" automatic phonograph mechanism. And still at 12 selections. What they lacked in technology they tried to make up with original design. The 5¢/10¢ coin chutes at the top are as modern as the age, as are the Deco lines. But Mills was running out of their string with this old construction. Equipped for remote speakers. $1,010.

AMI SINGING TOWERS, 1939. SINGING TOWERS was a whole new idea in jukeboxes. Tall and commanding, it demanded play. 10 record, 20 selection machine that plays either side of a record as selected. Model 100 introduced the machine in 1939, with production exceeding the AMI STREAMLINER and continuing into 1941. Molded glass panels were used instead of plastic. Model 201 improved its workings, and the factory reconditioned Model STR-100 was available during the war years. $6,050.

Wurlitzer 61, 1939. Made for locations that didn't have the room to put in a full size standing machine, countertop automatic phonographs were produced by Wurlitzer and Rock-Ola. The 61 is one of the early designs by Wurlitzer's own Paul M. Fuller, regarded as the dean of jukebox designers. An earlier countertop is Model 51, although Model 61 was the first to utilize plastic. A 12 record, 12 selection machine. $4,015.

Wurlitzer 700, 1940. The distinctive Paul Fuller Wurlitzer look began to take shape in the floor machines of 1939 and 1940. Selections were moved up to 24 with the 700, with display color provided by colored light bulbs behind the "Italian onyx" plastic pilasters. Nickel, dime and quarter coin slides. Selections are by pushbutton, a novel modern idea. Merchants could color match the machine to their decorating by using the correct color bulbs behind the plastic. $4,520.

Wurlitzer 800, 1940. Here comes the bubble machine. 800 is the first jukebox to use colored bubble tubes, a string of vertical glass tubes in the center of the grille. The outside pilasters provide a continuous display of moving colored lights. 24 piano type keyboard selections, triple coin chutes. Much of what the jukebox would become was worked out on the 800. Operators called it the "Eight Hundred," name enough for its fame. $4,810.

AMI SINGING TOWERS, 1940. If the 1939 SINGING TOWERS is big, the 1941 model is twice as big, in size and play. With two stacked AMI 20 selection units it is a 40 selection machine, playing both sides of 20 records. Its nickname is "Hi-Boy." The introductory Model 302 was produced from 1940 until 1942. Factory reconditioned Model HB-302 machines were available during the World War II years 1942-1945. Speakers at the top projected the sound. $10,050.

Seeburg SYMPHONOLA CONCERT MASTER, 1940. In a year that introduced six SYMPHONOLA models--COLONEL, MAJOR, ENVOY, COMMANDER, CADET and CONCERT MASTER--the CONCERT MASTER reigned supreme as the line leader. In an attempt to match the allure of the Wurlitzer machines, it had changing lighting "Rainbo-Glo" illumination, particularly effective behind the Greek comedy and tragedy faces at the top. 20 records, 20 selections, three coin slides. Nickname was "Faces." $6,020.

Rock-Ola 40 MASTER MODEL 1401, 1940. Model proliferation had taken over the jukebox industry, and if the war hadn't intervened there is no telling how many new models would have appeared from each manufacturer every year. The Rock-Ola 40 series, for 1940, included eight MASTER and SUPER models, half in walnut cabinets and half in Rockolite, a mottled painted finish that looked marbleized. 20 records, 20 selections. Model 1401 is walnut, 1403 Rockolite. Strong bass. $2,610.

Wurlitzer 850 SUPER DELUXE VICTORY, 1941. Top of the line for Wurlitzer in 1941, the elegant Paul Fuller 850 design includes continuous movement of Polarized light behind the front glass. Most models had the "Peacock" glass, giving the machine its nickname. 24 record, 24 selection. A lesser model without the polaroid lighting effects had a "Tulip" glass, for its nickname, but it is a rarer machine. One of the most desirable and valuable Wurlitzer jukeboxes. $16,050.

Wurlitzer 750 VICTORY, 1941. A smaller machine than the 850 for more modest locations, or so they said. Yet few Wurlitzer, or any, jukeboxes look as good as a 750 lit up in a darkened room. It's a knockout, and a Paul Fuller classic. Model 750 has a mechanical keyboard selector, Model 750-E an electric one. 24 record, 24 selection. Typically Fuller, the action of the changer is visible through a large picture window. $6,520.

Wurlitzer 780 COLONIAL, 1941. One of three classic 24 selection jukeboxes from Wurlitzer in 1941, the 780 was rated as a machine with subdued lighting and rich period beauty for quieter locations. The real reason for the Fuller designed 780 and its non-plastic display all wooden cabinet was the rising cost of these high tech materials and the difficulty of obtaining them in national emergency years. Wurlitzer called it COLONIAL, but operators preferred "Wagon Wheel." $3,810.

Wurlitzer 950 VICTORY, 1941. Rated as the most desirable and valuable Wurlitzer jukebox of them all, and the reason may be its rarity. The machine was introduced in 1942, just as wartime restrictions on materials and coin machine production ultimately led to a stoppage of all work. A Fuller design, you can just see the lines of the fabulous postwar Model 1015 beginning to emerge. 24 record, 24 selection. The first Wurlitzer to use fluorescent lighting. $25,050.

Rock-Ola COMMANDO MODEL 1415, 1942. Rock-Ola introduced a series of tall, top speaker boxes in 1941 with the PREMIER MODEL 1413, followed by PRESIDENT MODEL 1414 early in 1942, culminating with COMMANDO MODELS 1415-1418 during the war years, making use of pot metal, wood and glass as non-essential wartime materials. These were troublesome machines as the substitute materials didn't hold up. The PRESIDENT is the classic of the line, the rarest and most valuable. $8,515.

Aerion CORONET MODEL 400, 1946. At least a dozen new jukebox makers started out after the war, each with a unique idea. The Aerion Manufacturing Corporation of Kansas City, Kansas (the former Aircraft Accessories Corporation) created a 24 selection machine that featured neon lighting in a highly stylized cabinet. It came in blonde, cherry or natural walnut cabinets, with the cherry model shown. Inexpensive, it was meant to compete at the low end. Nicknamed "Canned Ham." $1,410.

Wurlitzer 1015, 1946. An eternally appealing feature of the 1015 is its small theater of music, with the 24 selection mechanism doing its wonderful work behind glass right before your eyes. A decade of dancing teens were fascinated by its movements. Prewar jukeboxes were for listening and dancing young adults, but the postwar crowd was a lot younger and more enthusiastic. They poured money into jukeboxes, and everybody wanted a piece of that pie.

Mills Industries CONSTELLATION STANDARD MODEL 951, 1947. After finally dumping their "Ferris Wheel" automatic phonograph in 1939 for the more conventional and exceptional EMPRESS and THRONE OF MUSIC jukeboxes, after World War II Mills Industries (newly renamed during the war years) opted for a completely new machine. The case is all aluminum, with a massive 40 selections. CONSTELLATION MODEL 950 was introduced in 1946, MODEL 951 in 1947. $1,210.

Wurlitzer 1015, 1946. The most classic jukebox of all time, and the one best known throughout the world. The power of numbers, and advertising, made that possible. Nothing exceptional, the 1015 was the logical extension of Paul Fuller's design work. But the machine came out after four years of war with no new machines and met an enormous pent up demand. Well over 50,000 were produced, with sales driven by a national ad campaign. Fame came as "The Bubbler." $7,515.

Aerion FIESTA DELUXE TYPE 1207A, 1947. Scrambling for market position was endemic to the many new jukebox producers after WWII. Aerion leaped into the market with four feet; four different jukebox models. The FIESTA STANDARD TYPE 1207 was introduced in 1946, and was buggy. So it was freshened up and returned in 1947 as the FIESTA DELUXE TYPE 1207A. The lower front panel is a mosaic of colors that continually flickers by backlighting. 24 selections. $1,110.

Rock-Ola MODEL 1426, 1947. The first postwar Rock-Ola jukebox was the 20 selection MODEL 1422, a beautifully lighted "Cathedral style" machine second in preference to the Wurlitzer 1015. It came back in 1947 as MODEL 1426 with a new front grille and lighting format. These are beautiful machines, and are recognized as classics. Display includes revolving lights behind plastic whenever a record plays and a visible changer, both Wurlitzer 1015 features. $4,510.

Wurlitzer 1100, 1947. Paul Fuller's last design for Wurlitzer, and a complete break from the "Cathedral" look he fostered for almost a decade. The postwar teens needed a new look, and they got it in the 1100. It introduced the light touch Zenith "Cobra" tone arm to reduce record wear. 24 record, 24 selection. 1100 did retain one classic Wurlitzer hallmark; the visible theater of record changing and playing. Nicknames: "Bullet," "Bomber Nose" and "Turret Top." $4,210.

Rock-Ola MODEL 1428, 1948. The basic Rock-Ola postwar jukebox design came back again in 1948 as MODEL 1428, this time sporting a luminescent plastic top with new side pillars and a patterned design replacement front in place of the earlier Deco grilles, all backlit by its "Magic-Glo" feature. But the clock was ticking on 20 and 24 selection machines as much newer and far more powerful machines were on the drawing boards, soon to arrive. $4,510.

Seeburg SYMPHONOLA P-148 BLONDE, 1948. Postwar Seeburg design followed much the same pattern. An upgraded version of the prewar 24 selection SYMPHONOLA mechanism was fitted into a cylindrical cabinet. 1-46W came out in wood in 1946; P-146 in metal soon after; P-147W in wood and P-147M in metal in 1947; and finally P-148M in 1948 and the P-148 BLONDE in a blonde finish metal cabinet. It generated several nicknames including "Washing Machine," and "Barrel," but "Trash Can" was the leader. $2,620.

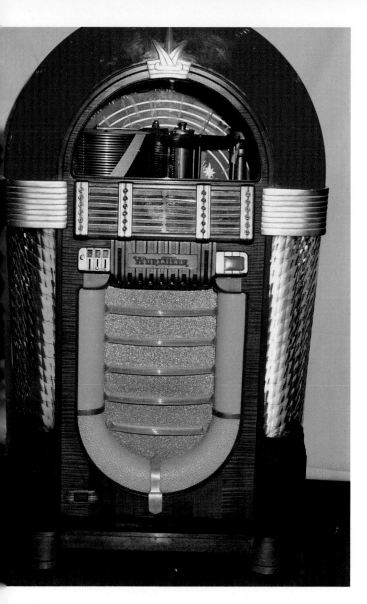

Ambassador THE AMBASSADOR, 1948. The staggeringly high population numbers of the Wurlitzer 1015 led to its acceptance as obsolete in a few years, particularly when its bubbles wore out. So kits were developed to change its looks. The leader was Ambassador, Inc. of Kansas City, Missouri with their THE AMBASSADOR revamp. Revolving lights and bubble tubes were replaced with static display materials. That was okay then. But the original 1015 doesn't look obsolete now. $10,100.

AMI MODEL C-20, 1949. Grand Rapids, Michigan, may have been off the beaten path, but that didn't deter AMI from making a dash for sales dominance with dramatic postwar jukes. MODEL A of 1946 was called "Mother Of Plastic" it was so laden with design. MODEL B of 1948 tempered it somewhat, and MODEL C-20 (for 20 record, 40 selection) was more conservative. It has a nifty feature. A mirror in the top reflects the changing and playing much like a Wurlitzer theater. $2,410.

Wurlitzer Automatic JUKE BOX, 1949. Wherever Americans went during WWII led to postwar jukebox fads, resulting in offshore producers in the U. K., Europe and elsewhere. In Australia it was learned that the Wurlitzer name was not trademarked. So a group formed Wurlitzer Automatic Phonograph Company of Australia Pty. Ltd. in Melbourne, and produced two machines. Their first, JUKE BOX, is a revamp of the Wurlitzer Model 600 of 1939. Wurlitzer sued but lost. $5,515.

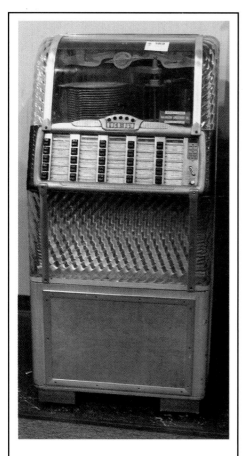

Rock-Ola ROCKET 78 MODEL 1432, 1950. Rock-Ola broke away from their classic styling with the ROCKET in much the same way Wurlitzer did with their Model 1100. It was promoted as a 1951 model, upping the 78 rpm play to 50 selections playing 25 records. But all that was fruitless as Seeburg came out with a 100 selection machine in 1948 and put the whole industry on notice. And things were about to change drastically with the coming of 45 rpm records. A ROCKET 45 model resulted. $2,610.

Chicago Coin BAND BOX, 1950. Smelling big money to be made in the jukebox business, Chicago Coin Machine Company of Chicago, primarily a pinball and sport themed arcade machine maker, entered the market. Their first project was BAND BOX, an animated jukebox and hideaway remote speaker. The curtain opens and the small figures appear to play when music is selected. The 1950 model has the copy "Strike Up The Band." Modified 1951 version says "Play Hit Tune Of The Week." $3,210.

Wurlitzer 1600 DELUXE, 1953. Sad days at North Tonawanda! Paul Fuller was gone, Wurlitzer management was screwed up, and Wurlitzers looked like Seeburgs. They copied the designs of their competition, a sure sign of trouble ahead. The 1600 was the dying gasp of 78 rpm records, with changeover kits available for 45 or 33-1/3 rpm. Whatever way, 48 selections. 1600 DELUXE is walnut. 1600 CUSTOM was available in a vinyl plastic finish in blue, blonde, red or mahogany. $2,610.

AMI MODEL E-120, 1953. Not to be outdone, AMI went for bigger selection numbers, covering all bases with three models of each machine. The "E" series was produced in E-40 (40 selections 78 rpm), E-80 (80 selections 45 rpm) and E-120 (120 selections 45 rpm) models. They look a lot alike, with great differences in the selection panels holding the title strips. A plus feature of the "E" is its capability to promote a location, with "B & B Bar" inside this glass. $1,600.

Seeburg SELECT-O-MATIC M-100C, 1952. The machine that changed the jukebox world. While others were fiddling with 20, 24, 40 and 50 play machines, Seeburg introduced the SELECT-O-MATIC M-100A in 1948. The "100" in its name comes from 100 78 rpm selections. M-100B of 1950 made that 100 45 rpm selections, and the 78 record was dead. The M-100C of 1952 added rotating color side pilasters in plastic, and 100 selections of 45 rpm became an industry standard. $2,500.

Ristaucrat RISTAUCRAT 45, 1950. The new 45 rpm records were a blessing of postwar technology, and took off in rapid popularity as inexpensive record players hit the home market. Moving into paying jukebox territory took a little longer. Some producers leapt right in, including Ristaucrat, Inc. of Kaukauna, Wisconsin, an old line jukebox maker. The RISTAUCRAT 45 played a stack of twelve 45s in sequence. A restyled selective S-45 model came out the next year. $400.

Seeburg SELECT-O-MATIC HF-100R, 1954. Another fabulous cabinet from Seeburg. This is the fabled mid-'50s styling that is sometimes known as "Detroit Deco." You'd be perfectly comfortable if this elaborate chromed facing was on the front of a '50s car. The "HF" stands for high fidelity, a sound reproduction catchword of the mid-'50s. HF-100R has a 25 watt high fidelity amplifier. Seeburg said the model had "Full-Spectrum High Fidelity and Omni-Directional sound." $2,600.

AMI MODEL F-120, 1954. Continuing with their triple threat format, AMI offered three models of the MODEL F: F-40 (40 selections 78 rpm), F-80 (80 selections 45 rpm) and F-120 (120 selections 45 rpm). This large, boxy jukebox looks different than the products of any other maker, with the exception of Bal/Ami in Great Britain, who made it under license. The rarest version is MODEL F-40 as only a few were made at the end of the 78 rpm era. $1,200.

Chicago Coin HIT PARADE, 1951. Chicago Coin even made a jukebox, a small 45 rpm player along the lines of the RISTAUCRAT 45, except it took nickel, dime and quarter and offered selectability. Inside is lighted with side and back graphics to create a miniature theater. 10 record, 10 selection. They turned to RCA, the maker of the 45 record, for a reliable changer. Nice, entry level inexpensive jukebox collectible, worth double when accompanied by its original floor stand. $800.

AMI MODEL H-120, 1957. In a dramatic departure in design, the AMI "H" series is much smaller than the big box E, F and G machines that preceded it. Even the selection count was altered, with H-100, H-120 and H-200 (for 100, 120 and 200 selections in 45 rpm only) models produced. Once again, the differences are apparent in the handling of the title strips. The H-200 model was known as the SHOWBOX, and has the number "200" next to the AMI logo on the front. $1,600.

Seeburg SELECT-O-MATIC KD-100, 1957. 200 selections! Common stuff by the end of the '50s, with smaller, lighter and more durable 45 rpm records giving new life to jukeboxes. Records lasted a lot longer. But filling a 100 record, 200 selection box could also be an expensive proposition for any operator. A collector, too. Seeburg's KD-200 is the ultimate in "Detroit Deco," with what appear to be automotive tailfins with red stop lights. Printed circuit board. $2,500.

Seeburg SELECT-O-MATIC AQ-160S, 1960. Where has all the styling gone? The Seeburg "AQ" series added the option of playing 33-1/3 and 45 rpm records in both 100 and 160 selection models. But the cabinets are awful. It's as if the design track followed the popularity trail; they both dipped. The AQ-160S has an interesting gimmick in its "Artist Of The Week" panel in its marquee. The operator can slip in a promotional flyer or album cover to push a name. $1,200.

Wurlitzer 2500-S, 1961. Stereo is a major part of the action by 1961, with the 200 selection 2500-S carrying the large copy "Stereophonic Music" across its header above the viewing window. 2500 came in both mono and stereo models, with 2504 a 104 selection and 2510 a 100 selection machine. Even with the 200 selections, the wide fan of the title strips make for easy reading. But new technology couldn't counteract apathy as TV took over the entertainment industry. $1,200.

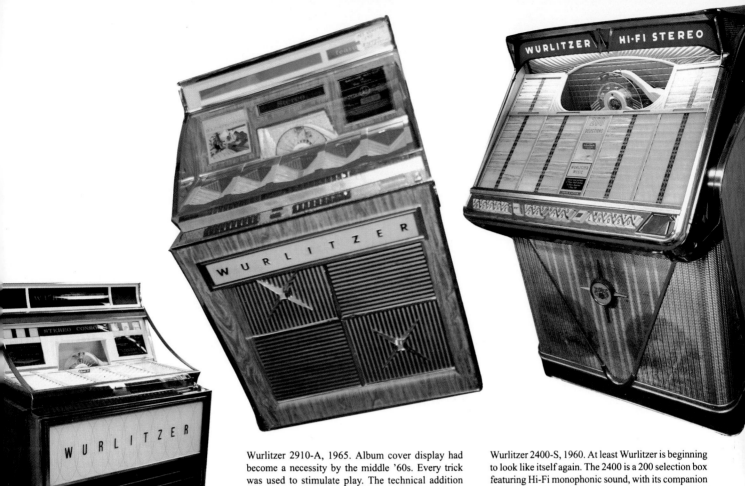

Wurlitzer 2910-A, 1965. Album cover display had become a necessity by the middle '60s. Every trick was used to stimulate play. The technical addition in the Wurlitzer 2900 series was a solid state amplifier. Model 2900 was a 200 selection machine; 2910 was 100 selection. To go beyond the normal, the Model 2910-A was available in a blonde cabinet. Increased title strip display and promotional graphics continued to chip away at the viewing window, by now only a peep hole. $500.

Wurlitzer 2400-S, 1960. At least Wurlitzer is beginning to look like itself again. The 2400 is a 200 selection box featuring Hi-Fi monophonic sound, with its companion model 2400-S substituting Hi-Fi stereophonic sound. It came both ways, only you paid more for the stereo. To show the difference, the words "HI-FI STEREO" are loudly proclaimed across the lighted top panel, above the window. The greater number of selections reduces the size of the viewing window. $650.

Wurlitzer 3000 STEREO CONSOLE, 1966. The top display area of the Wurlitzer 2900 was carried over to the 3000, although the limiting side walls were eliminated and the up-and-down title strips were flattened out for easy reading. 3000 is a 200 selection machine, and 3010 is 100. The machine also handles Top Tunes and the little LP albums with three selections on a side. The curved glass top lifts in one piece for fast title strip changing. $600.

Seeburg SELECT-O-MATIC LPC-1 CONSOLE, 1962. As TV grew in stature and gained in audience people went out less, and soon there were fewer locations to run jukeboxes. Just as bad, the market was so saturated with 45 and 33-1/3 rpm machines in good condition there was little need for replacement. Staid cabinet design offered promotional opportunities such as multiple display of album covers and leading artists (this LP CONSOLE has Elvis and The Beatles). 80 records. $800.

Wurlitzer 3700 AMERICANA, 1973. The AMERICANA name was too good to drop, particularly for export machines, so Wurlitzer brought it back in 1973. Jukebox styling had reached its nadir by the middle '70s, with many looking a lot alike. The Wurlitzer 3700 is a 200 selection jukebox, and the 3710 offered 100 selections. By this time Wurlitzer was in real trouble, and the future looked grim. It looked that way for all jukebox makers. $400.

Wurlitzer 3100 AMERICANA, 1967. The AMERICANA looks like style change for the sake of change, with the production numbers running ever lower. The jukeboxes from the 1960s and '70s, while they are technically exceptional, do not command significant collector prices. The result is the availability of good machines at rather modest cost, perfect if you are looking for controlled music in a basement bar or rec room. 3100 is a 200 selection machine while 3110 provides 100. $500.

Wurlitzer 3000 STEREO CONSOLE, 1966. Over the years the modern jukebox had evolved into an extraordinarily complex device. An inside look at the Wurlitzer 3000 from the back shows the maze of wires, the record carousel and the vertical playing turntable. Wurlitzer even tried to promote the 3000 as a Discotheque machine with matched speakers and special Disco Music Library on records, quickly reconvertible to pay-for-play. But the public wasn't buying. $600.

Seeburg SMC-1 DISCO 160, 1978. Collectible jukeboxes seemingly ended with the Wurlitzer NOSTALGIA, yet the business goes on. The Seeburg Corporation, a new entity with a grand old name, created the DISCO 160 in 1978, a pioneer machine with the "New Look" for jukeboxes. 160 selections, 80 records 45 rpm. It features a "Cupolium Sphere" that lets you see the changer and player in action. Newer machines play CD discs. They will ultimately all be collectibles. $650.

Seeburg SPS2 MATADOR, 1973. If you like the effect of continually rolling purple polka dots, the MATADOR is for you. Machines like this were designed to glow in dark barrooms and restaurants while they provided soft background music. The selector panel in the center works like a calculator, allowing you to punch in the three digits of a selection, thus eliminating the ever longer rows of pushbuttons that had plagued jukebox design since the late 1930s. $700.

Wurlitzer 1050 NOSTALGIA, 1973. The old adage once again. When you look to the past for glory, there probably isn't much in your future. The NOSTALGIA was a blatant copy of the classic Paul Fuller styling of the immortal Wurlitzer 1015 of 1946. It didn't catch on, and cases of the machines sat around for years until the collector market began to pick them up. The next year, after making one more model, the AMERICANA 3800 series, American Wurlitzer folded, the name surviving on jukeboxes made in Germany. $3,200.

Pinball Games

Pinball is a unique form of mechanical entertainment, and is one of only two original forms of amusement machines to come out of the American coin-op industry. The 1880s and 1890s produced coin operated versions of the vending machines, scales, strength testers, phonographs, and the saloon bartop dice games and the like that had existed prior to the new mechanical art. But only the automatic payout slot machine (there had been nothing like it before coin-ops came along in the 1880s) and the pinball game were truly original creations that relied on the function of coin control. Imagine that! A glass covered marble game that you couldn't control with your fingers, an 1890s development. These machines were different.

The idea of adding coin control seems to have happened in the late 1890s, with a few commercialized games available in the early 1900s. But they didn't catch on.

It wasn't until the 1930s that bagatelle was once again put under coin control, and this time it stuck, probably because it offered cheap entertainment at a time when hardly anybody had money to spend. A few games had been made between 1928 and 1931, but they died aborning. Until Automatic Industries in Youngstown, Ohio, made a game called WHIFFLE. Whew! Did that baby take off. They were produced in the tens of thousands of units by the end of 1931, and a new game idea had been born. Except a company in Chicago, making a game called

Inwersen HOME PARLOR KEEPS, 1890. 19th century bagatelles and marble games shot marbles onto a tilted playfield allowing the ball to move downward to enter scoring holes. HOME PARLOR KEEPS is a small tabletop game based on a larger saloon gambling table. The ball goes into the center hole and comes out in one of the score troughs below. Maker is E. G. Inwersen Manufacturing Company of Lyons, a Mississippi river town gobbled up by Clinton, Iowa, in the early 1900s. $250.

Caille Schiemer LITTLE MANHATTAN, February 1901. A saloon gambling game called MANHATTAN was miniaturized by the Caille Schiemer Company in Detroit, Michigan, as the LITTLE MANHATTAN. A competitive two player game, shooters are on each side of the playfield. When designer Adolph A. Caille joined his brother Art to form The Caille Brothers Company in July 1901 production continued. This is the four scoring pocket model, matching playfield holes. One pocket model also made. $900.

Pinball came out of a tilted table game called bagatelle that originated in France about the time of the American Revolution. It evolved into both a child's toy and, in much larger versions, a saloon gambling table in which marbles were shot onto a playfield by a shooter, curving around an arch at the top of the playfield and coming downfield to make scoring holes or other playfield features, such as bells, spinning pointers and other gimmicks.

WHOOPEE at the same time, thought that *they* had started the boom. The boom hit hard in 1932, with the new "Depression Baby" marble tables becoming one of the hottest fads in the country.

What really happened was that a form of amusement that is still with us had hit the boards just at the time it was needed as cheap Great Depression entertainment, a game for a penny.

The advertising for the games began to leak into the trade publications, such as *The Billboard*, and the even more esoteric publications such as *The Coin Machine Journal* and *Automatic Age*. Literally unknown to the trade in 1931, by the time of the 1932 coin machine show in Chicago there were over 60 manufacturers of pinball games (which were called marble games then. The word "Pinball" came a few years later as the game format matured.) on the floor, with over a hundred different games being exhibited. The boom was on, never to end!

The big makers were D. Gottlieb & Company in Chicago (which is still around as Premier Technology), Bally Manufacturing Company (a pinball game called BALLYHOO *started* the company), Mills Novelty, O. D. Jennings, and Rock-Ola. A young David C. Rockola jumped right into the ring, and for about half a dozen years was one of the major producers of the game. His classic was a 1933 game called JIG SAW. The game was made to commemorate the Chicago World's Fair of that year, and was often called the WORLD'S FAIR JIG SAW, although JIG SAW is its true name. It is one of the highest population pinball games ever made, with a production of about 70,000 (production of a typical game today goes to about 2,500 units). In spite of that, it is one of the most valuable of the vintage games.

Strangely, the jig saw as a pinball feature was tried only twice after that. First, in another game by Rock-Ola in late 1937 called JIG JOY, in which the jig saw puzzle was simulated in the backglass. Second, the Williams Manufacturing Company (Literally a newcomer to the business. The firm was founded in 1946 as a revamper, taking old games and making them new again. It evolved into the world dominant electronic giant of today) made a game called JIG SAW in November 1957 that did the same thing, only not as well.

But the honor of being the only game that has actual cut-out jig saw puzzle parts that assembles into the completed puzzle when the proper holes are made remains with the original Rock-Ola JIG SAW. It was a classic idea that survived the years to become, in collector terms, hot. There are a lot of other classic pinball games that have very distinct and unique features, with the games avidly collected by a rapidly growing segment of the coin-op collector population. Once collectors caught on to the fact that they could actually own a commercial game (and once the distributors caught on to the fact that there was a lucrative home market to sell) pinball fan clubs and annual events sprouted up all over the country.

Pinballs are the coin machines that most people are familiar with, and there are probably more pinball games in different private homes than any other class of coin machine. So why did it take so long to become a viable collectible? It's a matter of attention span. Pinball people are, well, different. As collectors they don't see themselves as coin-op enthusiasts. They are pinball enthusiasts, and the fact that their machines work by a coin (although many of the "home" machines have been converted over to free play by pushing a button or tripping a lever, which in

Caille Bros. LOG CABIN, January 1904. The first commercially successful pinball game, the coin operated Caille LOG CABIN wasn't recognized as such when it came out. It was seen as a trade stimulator that paid off in drinks or cigars for high scores, with its colored marbles doubling their count. The initial "Square Top," entered production in Detroit in 1901. The "Round Top" came along in 1904, and was sold until Prohibition. Vertical plunger kicks ball. All models rare and valuable. $300.

Automatic Industries BABY WHIFFLE, November 1931. Bagatelles survived well into the historic pinball period, and plastic versions can still be purchased. Automatic Industries, Inc., of Youngstown, Ohio, the claimed inventors of pinball with their coin-op WHIFFLE table game of June 1931, made bagatelles for both home and commercial use. BABY WHIFFLE contains features of the coin machine, including the ball lift (wheel at right) and ball return shuttle (lever at lower center). $400.

itself provokes major disdain among coin-ops collectors) is of little matter to them. If anything, needing a coin is an annoyance. They want to play, not pay. Collectors tend to specialize, by manufacturer (Gottlieb is hot!), by time period (the '50s is very popular) or by formats. But if you are a pinball collector, or have a machine or two, you are in for some rude shocks. As a general rule they aren't worth much. Some games are valuable to be sure such as that Rock-Ola JIG SAW and its companion Rock-Ola WORLD SERIES. These go at anywhere from $600 to $1,750 depending on the model and condition. But most other games from that period have a hard time topping out at $200, provided they are in prime shape. It's even worse with later machines. A lot of people paid a lot of money to get a game in the house, and when it finally came time to sell or get rid of it they found out that they almost had to pay someone to take the game away.

That's the down side. There are up sides, and they top the down sides by a wide margin. No coin-op is more fun to play! None! And if you are lucky enough to get a game that you grew up with in grade school, high school or college, it's a fountain of youth for you as you try to make the same holes and scores you did as a kid. That's glory! To those true devotees, they'll never give up their game. The furniture goes first. So how can you put a value on that?

Game-wise, there are 5 major pinball periods. The mechanical countertops and early legged machines (circa 1931-1934) are marvels of ingenuity and often a lot of fun to play. But they aren't worth much, maybe $75 to $225 at the top end in prime condition from a dealer. Figure the auction value or selling price at half that.

The battery powered electrics and early payouts (circa 1934-1936) add to that marvel of ingenuity, and many of these games are classic collectibles. And that makes them worth more. A Bally ROCKET that pays out well and doesn't have a trashed cabinet is valued at around $350-$750, depending on where you find it. But most of these games are in the $150-$250 value range. It was during this period that some of the most dramatic changes took place in the game. The unique part of this phase of pinball development is that not all of the new ideas came out of Chicago. Sure, 99-99/100ths did. But there are a few tantalizing auslanders that make the study of pinball interesting. In the early days of the game—1932 to around 1935—many of the game ideas came out of the West Coast, from Southern California to the tip of the State of Washington, and some producers even made their games there. Rube Gross of Seattle is one. He started late in 1934, and finally folded his game tent a few years later. In the meantime he created many very interesting games. FURY was his best seller, created in 1935. The name came from the highly original little kickers all around the playfield that whacked the ball all over the place once it was in play. That's why the batteries were needed. Value? Hard to say, as it hasn't been cataloged in a price guide. The important thing is, you have to know the games.

By 1937 the electromechanicals with plastic bumpers and lighted backglasses had entered the scene, and they revolutionized the game. Machines in this time frame generally are in the $150 - $250 value area, although the Bally BUMPER goes for $450. This trend carried right through World War II. The war-time revamps are worth the real money. SLAP THE JAPS, a rebuilt regular game, is worth $700 easy, provided the glass and playfield are prime. Then the Gottlieb HUMPTY DUMPTY came along in October 1947. It added a whole new element to game play. Gottlieb called them "flipper bumpers," but they later became known as flippers. That was the beginning of the modern game. It has only gotten racier and faster since then.

A Golden Age of pinballs began with that single improvement, with flipper "Woodrail" (1950s) and "Steelrail" (1960s) games moving into the $150-$250 value class, although that revolutionary HUMPTY DUMPTY is worth two to three times as much, and is going up, up, up! By the 1970s the modern game with ramps, pop bumpers, skill shots and fast action led straight to the unbelievable solid state and digital display pinball world you see under glass today. If you haven't played a pinball game in the last 5 or 10 years, try a new one. It'll scare you. But watch a kid play and control the action. It's magic!

Hot pin lovers will pay around $1,850 for a recent game just off location (they cost about $3,500 new), but can't get more than 10% of that back if they hold for 3 or 4 years and try to sell. There are some exceptions. The Bally CAPT. FANTASTIC "Celebrity Series" game of 1975 still sells for over $1,000, whereas the Bally DOLLY PARTON celebrity game of 1978 is such a dog to play it's hard to give one away.

Buckley FAVORITE, April 1932. Following the introduction of the tabletop Gottlieb BAFFLE BALL in November 1931 and the Bally BALLYHOO in January 1932, the Depression born pinball boom was off and running. These two landmark games established format and size for another two years, with BALLYHOO the leader and most copied. Buckley Manufacturing Company in Chicago made a duplicate of the BALLYHOO with their own horse race playfield as FAVORITE. 10 balls 5¢. $250.

Last notes. The "cosmetics" of a pinball game are the most important value and collectibility factor. A good game with a bad backglass and a worn playfield isn't a good game. A mediocre game with good glass and playfield is worth more. If the game doesn't work, that isn't as important as bad cosmetics. You can fix the mechanical stuff, but you usually can't replace the original glass or painted cabinet and playfield.

In short, for fun pick up a pin. Or two or three. A lot of the major pin collectors have added additions to their homes to incorporate the games, or even built a separate building to house them and their wonderful noisy action. But as an investment, stay away from pinballs unless you really know what you're dealing with.

Gottlieb FIVE STAR FINAL JUNIOR, May 1932. BAFFLE BALL and subsequent games made D. Gottlieb & Company of Chicago the leading pinball producer of its time, a position it held for almost half a century. FIVE STAR FINAL was a popular mid-1932 entry, introduced in a small size. When a larger table version on legs was introduced as FIVE STAR FINAL SENIOR in June, the original became "JUNIOR." Typically 7 ball 1¢. Playfield widely copied. This is a non-coin model for home use. $300.

Hutchison SCRAM, May 1932. Balls rolling uphill? That's what seems to be happening in SCRAM, produced by the Hutchison Engineering Company of Nashville, Tennessee. The playfield slants away from the player, with the scoring balls held captive until hit one at a time by a larger ball propelled by the shooter, coming around the arch below to go into scoring pockets. A. B. T. Manufacturing Company, Inc. and J. P. Seeburg Corporation, both of Chicago, marketed this game. $400.

Exhibit Supply PLAY BALL, August 1932. An old line arcade machine and counter game maker, the Exhibit Supply Company of Chicago longed to get into the booming pinball business. After a few unsuccessful trys they finally clicked with a BALLYHOO size game called PLAY BALL with graphics based on a counter game they made in 1926. It sold as a dedicated game and as an insert board available separately to upgrade old Bally games. Exhibit Supply became a major pinball maker of the 1930s and '40s. $300.

Houston Showcase SWEETHEART DOUBLE, March 1933. The Lone Star State was an early hotbed of pinball design, and the home of the two-player game. The Houston Showcase & Manufacturing Company of Houston, Texas, facing hard times with its showcase business in the Depression, turned to pinballs. In the summer of 1932 they made a game called WILD CAT, followed by SWEETHEART in 1933. Then they made double playfield versions of both games, and the 2-player was born. $350.

National Pin THE PILOT, November 1932. In 1932 people still ran out of their houses to see an airplane when they heard one. It was usually a biplane; the monoplane was still a rarity. Which makes the graphics on THE PILOT very advanced for its day, resulting in a wonderful period piece. Maker was National Pin Games Manufacturing Company of Detroit, Michigan. It's a skill game. You want to get the ball in the top propeller, controlled by the center knob on the cabinet. $300.

Rock-Ola WINGS, February 1933. Getting into pinball manufacturing in the summer of 1932, Rock-Ola Manufacturing Corporation of Chicago almost lost their shirt and their business with a relatively unsuccessful game. They waited almost a year to come back with WINGS, an exciting game with a spinning playfield. When you push the ball lift it spins, and you're shooting at a moving target. It was the first such device in pinball. $350.

94

Automatic Games FIFTY GRAND, 1933. Conventional history has it that light-up features, a backglass and automatic scoring didn't come along until 1934 and 1935, and even then they were embryonic. Yet here is a game from 1933 that has it all. Historical oddities like this are of great value to serious collectors. Maker is Automatic Games Company of Inglewood, California who got started in 1932 making a series of arcade diggers called PILE DRIVER, STEAM SHOVEL and SCOOP. $600.

Rock-Ola JIG SAW, August 1933. An incredible exercise in ball management and feature manipulation, all done with hidden wires and hinges. There is no electricity in this game, yet it pieces together a jig saw puzzle of the Chicago World's Fair of 1933 when the proper holes are made. Dropped balls push levers, and the pieces are flipped in place. Fast becoming the most desirable vintage pinball game of the pre-flipper era, with its value continually climbing. $1,600.

Chicago Coin LELAND STANDARD, December 1933. The Chicago Coin Machine Company was formed to market pinball games without production facilities. So the Chicago firm subcontracted games to two brothers named Stoner with a woodworking shop in Aurora, Illinois. The LELAND was called the "Lucky New Horse Shoe Game" for the casting at the playfield top that gave the ball a fast spin. Larger table model was called LELAND DELUXE SENIOR, a small countertop LELAND DELUXE JUNIOR. $250.

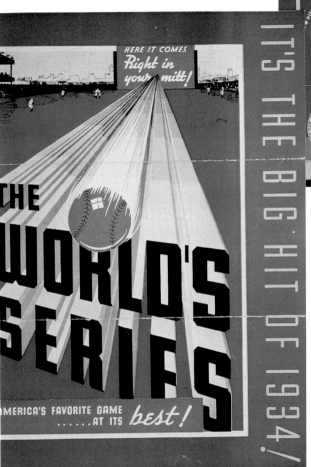

Rock-Ola WORLD'S SERIES, May 1934. Another mechanical masterpiece from Rock-Ola. If the ball makes a base hit hole it lands in the baseball diamond. Another base hit and the diamond spins the ball to the next base. Home run and they all spin home. Rock-Ola also elevated the lowly pinball flyer from a black & white sheet to a multi-page multi-color art form, becoming an industry standard for a number of years. The flyer was so good, others copied the game before it was even released. $250.

Rock-Ola ARMY NAVY, December 1934. When Bally came out with a clever action game called PENNANT in October 1933, football was the game to replicate. Rock-Ola's all mechanical ARMY NAVY was truly ingenious and an infinitely more complex device that moved a football across the lower playfield on a small chain conveyor depending on the holes and plays made in the upper playfield, also keeping score. This was the apex of the mechanical game as electricity was taking over. $550.

Chicago Coin BEAM LITE, April 1935. Loosely imitating the Stoner BEACON, the first lighted playfield pingame, the Chicago Coin BEAM LITE sported 13 light bulbs that were kept beaming in play by 8 batteries. They were colored by placing tinted pharmaceutical gelatin capsules over the bulbs which also protected them from the fast moving balls. Plug-in games had not yet arrived, but they were coming along with transformers to step down the voltage. $350.

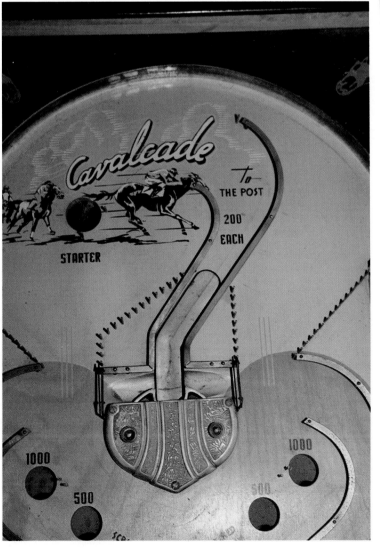

Genco BASE BALL, June 1935. Three Gensburg brothers formed Genco, Inc. in Chicago in 1930 (a fourth founded Chicago Coin) to make copycat pinball games to sell at lower cost. They became a major original game producer. Sometimes. Other times their copycatting tendencies were most evident. BASE BALL looked enough like the Rock-Ola WORLD'S SERIES to lead to a lawsuit, which ended in a draw. For one thing, BASE BALL is electric. Early backglass, unlighted. $500.

Stoner CAVALCADE, April 1935. Once they saw how many games Chicago Coin was selling, Stoner brothers Ted and Harry decided that was the business for them. Stoner Manufacturing Corporation, Aurora, Illinois, was formed October 1933, producing clever games until World War II, then going into vending. In CAVALCADE you fill up the center post hole with balls, and then make the electrical kicker "Starter" hole, which makes the stacked balls shoot out left and right to score. $350.

Pacific LITE-A-LINE, November 1934. Pacific Amusement Manufacturing Company of Los Angeles is credited with adding electricity to pinball for a bell ringing device on CONTACT in late 1933. They became known for their battery driven electrical games. LITE-A-LINE is rated as pinball's first light-up game. Three bingo card style back panels light up numbers when they are made in the roulette drum. This 1934 game was still in commercial use on new legs until 1954. $450.

Bally JUMBO, August 1935. Bally Manufacturing Company was a very successful pinball producer, and led the industry in payout games. The idea was to make certain nearly impossible playfield scores, and out would come money. Payout pins are highly desirable to collectors. JUMBO gives you only one ball for a nickel, so it has to count. You could win up to $2. The winning holes are so protected one wonders if they were ever hit. Which is the whole idea. $450.

Rube Gross FURY, 1935. The Pacific Northwest was another pinball hotbed, with distributors and designers making original games. These are generally rare, and elevated in value. Rube Gross & Company, Inc, a Seattle, Washington, punch board and slot distributor, made TORPEDO in 1934. The colorful FURY, made in 1935, incorporated all the latest technology, including a series of solenoid kickers that manipulated the balls before shooting them up the playfield once again. $400.

Genco SPIT FIRE, August 1935. A Genco original. They even got a patent on the idea. Electricity allows the game to have all sorts of new features, including kickers that send the balls moving fast along controlled runs. When the top holes are made the balls are kicked underneath, and come flashing down elevated wire tracks to spin around and go shooting up the playfield. A red light flashes when they come out of the holes. Good condition games like this are highly sought. $350.

Gottlieb SUNSHINE DERBY, January 1936. By 1936 pinball had acquired a backglass. It wasn't very tall, nor did it have many lighting features. But it was a part of the game. SUNSHINE DERBY has additional charm; it's a payout pin, one of the few made by Gottlieb. Picking up on the Bally format, this is a one ball machine. It also shows off some major advances in design, such as the painted cabinet, woodrail sides and sturdy legs. $350.

Pacific PAMCO PALOOKA, February 1936. An advanced development of the Pacific LITE-A-LINE of November 1934, the game works on the same principle. But now the display is in a single backlighted colored panel, and there are six coin slides to pick your color and shoot. Reason for the higher odds? It's a payout, working the same way as the old turn-of-the-century 6-way floor slot machines. Pacific Manufacturing Company Inc. was so successful, they moved to Chicago. $400.

Jennings WALL STREET, November 1935. A major slot maker, O. D. Jennings & Company of Chicago made pinballs to fill out their line. They combined their Bell machine and pinball skills into a payout game that spits coins when special playfield features are made. The hunting scene SPORTSMAN came out in 1934; IMPROVED SPORTSMAN in 1935. The WALL STREET version has graphics of LaSalle Street in Chicago, the financial center. Great collectible for a broker. $1,000.

Genco CHEER LEADER, October 1935. Another Genco original, they called it the "only lite-up football game!" If the ball goes into the channel at the left, it gets kicked up a notch or two or three at a time as holes are made in the playfield at the right by other balls. The backglass is lighted and shows the progress of the ball. The game plugs in and is controlled by a transformer. Genco became a leader in lighted backglass games. $300.

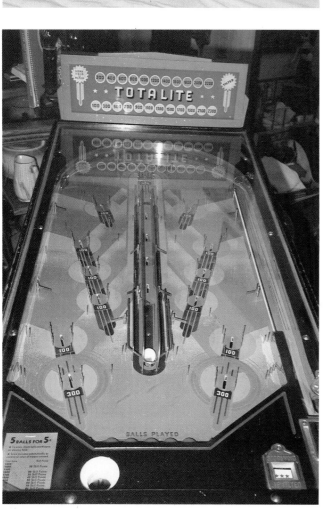

Rock-Ola TOTALITE, May 1936. Never one to be left out of advanced developments, Rock-Ola scooped the industry by coming out with the first lighted backglass game with automatically advanced scoring. One of the key features of the modern pinball game was now in place. This form of scoring remained virtually unchanged until the mid-'50s. To add to the excitement the ball was shot along an elevated runway if it made the TOTALITE kicker hole at the top of the playfield. $450.

Bally BALLY BOOSTER, April 1937. Directly competitive to the Gottlieb ELECTRIC SCOREBOARD, even coming out the same month, BALLY BOOSTER was an advanced development of BUMPER. The light-up scoring did Gottlieb one better by not only providing a tally of the runs, but also running the bases on the backglass with lights. There is also a "batter" at the bottom of the playfield in the form of a covered kicker that whacks the ball upfield in a hurry. $500.

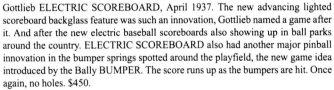

Gottlieb ELECTRIC SCOREBOARD, April 1937. The new advancing lighted scoreboard backglass feature was such an innovation, Gottlieb named a game after it. And after the new electric baseball scoreboards also showing up in ball parks around the country. ELECTRIC SCOREBOARD also had another major pinball innovation in the bumper springs spotted around the playfield, the new game idea introduced by the Bally BUMPER. The score runs up as the bumpers are hit. Once again, no holes. $450.

Baker ON DECK, May 1940. Pinball backglass grew tall, with backlighting of advancing scoring and multiple graphics features. New manufacturers didn't enter the market as fast as the early '30s, but it still happened. Baker Novelty Company in Chicago, a slot and console revamper, wanted to be in the business. So they had Chicago Coin make their games under private label. ON DECK lighted nautical ladies, and shows the beginning of inflated scoring numbers. $350.

101

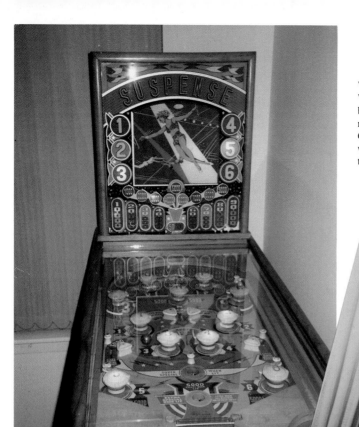

Williams SUSPENSE, February 1946. Pinball wasn't produced during the World War II years. Older games were upgraded with new names, playfields and backglasses. Harry E. Williams, the dean of pinball designers, was a wartime revamper. Late in 1944 he formed his own Williams Manufacturing Company in Chicago. His first game from the ground up was SUSPENSE, made soon after the war, with premier pinball artist George Molentin doing the backglass. Lighted plastic bumper caps. $300.

Keeney FIRE CRACKER, March 1937. The beginning forms of pinball's future are evident in the lines of the Keeney FIRE CRACKER, a mid-'30s game with an advancing score backglass. J. H. Keeney & Company of Chicago got into pin games early as a subcontractor of the Gottlieb BAFFLE BALL in 1931, keeping pace with the industry over the years. Play action is based on the Bally BUMPER of January 1937, quickly copied by Keeney. It's all action; there are no holes on the playfield. $400.

Gottlieb NEW DAILY RACES, February 1947. Bally was noted for its large, console style one ball horse race themed payout pin machines. Gottlieb made just short of a duplicate copy of the Bally BALLY ENTRY/SPECIAL ENTRY of December 1946 and gave it a name they had used in the past: DAILY RACES. They added the small word "New" in the backglass to denote the difference. These games are very suspenseful. Gottlieb gambling machines tend to be rare. $500.

Exhibit Supply SMOKY, March 1947. The pinball designers at Exhibit Supply Company in Chicago were Harry Williams and Lyn Durant, both geniuses of the game. When they left early in WWII to form United in Chicago, with Williams going off on his own soon thereafter, Exhibit Supply lost its design team. With the war over and pinball's return, Exhibit Supply games were unimaginative. The play of SMOKY was the same as VANITIES of February 1947, the game before it. $250.

Genco TRIPLE ACTION, January 1948. First Genco pinball with flippers. Game designer Steve Kordek (who remains in the business at Williams well into the 1990s) had only one set to work with. So he put them at the bottom below a turn disc feature, and added to their kick so they could pop the ball to the top of the playfield to get another chance at the game's scoring bumpers. It became the desired and classic flipper position. Backglass by Roy Parker, a leading artist. $450.

Exhibit Supply TALLY HO, November 1947. The introduction of flippers on Gottleib's HUMPTY DUMPTY in October 1947 caught the industry by surprise. Other makers kept going with their flipperless games, and they quickly became a drag on the market. This was particularly true of Exhibit Supply whose games didn't change. TALLY HO looks a lot like SMOKY of the previous March, and even has a horse on the backglass. Many were upgraded by add-on flipper kits in 1948. $250.

Gottlieb HUMPTY DUMPTY, October 1947. The game that changed the game of pinball, making it an undeniable skill machine and leading to the fabulous machines of the present era. HUMPTY DUMPTY added flippers to the game. Gottlieb called them "Flipper Bumpers" and strung three of them along each side, top to bottom. They were very slow acting. But they galvanized play. Truly a historically significant pinball machine, and much sought after. $750.

United NEVADA, October 1947. When designers Harry Williams (game design) and Lyn Durant (electricity and producibility) left Exhibit Supply to revamp games during WWII they formed United Manufacturing Company in Chicago. Williams left in November 1944 to form his own company, and Durant carried on making original games. United games have a three rollover "Spell Name" feature which lights up the game name on the backglass in pieces when the bumpers are hit: NE-VA-DA. $250.

103

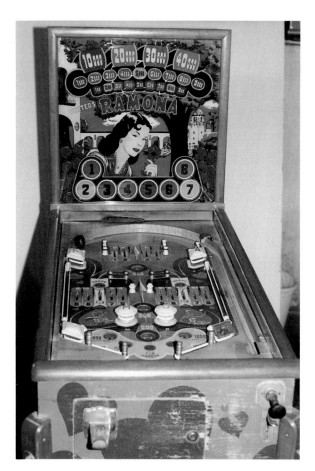

United RAMONA, February 1949. United resisted flippers, believing them to be a passing fad. They didn't add them until CARIBBEAN, May 1948, sticking them high up on the playfield, where they can be seen on RAMONA. RA-MO-NA was another spell name game with a basic United playfield and inflated backglass scoring. The two top bumpers spell RA, the center pop bumpers MO, and the bottom corner bumpers NA. $250.

Genco 400, October 1952. Vertical pinballs are rare, but a few were made. Genco made two. WHIZZ, a small countertop game (that may have led to the Japanese game of *Pachinko*), was made in October 1946. 400 was an overgrown WHIZZ in a floor cabinet. The game is very, very fast as gravity works against you all the time. But it is exciting and fun, and wonderful to watch as free games are racked up. Scoring is by drum, replacing advancing backglass lighting. $250.

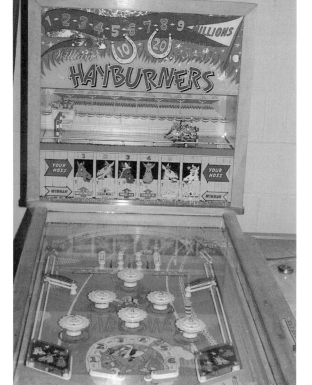

Gottlieb ARABIAN KNIGHTS, November 1953. The next game in the Gottlieb line, ARABIAN KNIGHTS is a classic of player involvement. Selector knob on front lets the player spot the 1, 2 or 3 holes that will accelerate scoring. The bumpers are silverized, made by laminating foil over the plastic tops. And the artwork is fetching. But the burden of scoring is still left up to the backglass. In classic format, two flippers at the bottom. $500.

Williams HAYBURNERS, June 1951. Pinball quickly evolved into a fast and exciting game that required skill and strategy, with feature after feature added (and just as often subtracted due to excessive production cost). HAYBURNERS offered an animated horse race in a deep backbox, with 3-dimensional horses lurching forward as thumper bumpers are made. Get a horse to cross the finish line with your first ball and you earn 20 free games. Flippers at bottom. $500.

Williams CLUB HOUSE, November 1958. Drum scoring starts to show up in general pinball, although the backglass still has a lot of the work to do. Two games are going on at once in CLUB HOUSE, with pinball bumper rollover scoring on the lighted backglass while a 21 blackjack game feature is counted on two drum reels behind a window in the center of the backbox. The 21 game dealer's score is unknown during the game, lit when over. Beat it and you get a replay. $500.

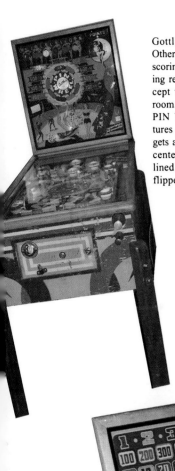

Gottlieb PIN WHEEL, October 1953. Other makers were slow to accept drum scoring, with advancing backglass lighting remaining the preferred solution. Except the backglass was running out of room. Backglass crowding is evident in PIN WHEEL, with a plethora of features jammed together hoping that each gets attention. Trap holes in the playfield center award replays when 3 balls are lined up, or 4 fill out the square. Four flippers across bottom. $450.

Gottlieb HI-DIVER, April 1959. 1950s pinball games are interesting and exciting to play. Many collectors specialize in '50s machines, particularly Gottlieb. A great innovation was backglass animation, with HI-DIVER known as one of the best. Divers jump into the pool in the backglass as the game progresses, coming out to jump in again. Artist Roy Parker added interesting touches. The lady lifeguard at lower right on the glass is reading "Super Men" magazine. $850.

Gottlieb QUEEN OF DIAMONDS, June 1959. Gottlieb was particularly noted for poker themed pinball games, providing an exceptional blend of playing pleasure, playfield and backglass graphics and their understanding of playing card scoring. The combination worked time and time again in an extended series of successful games, with QUEEN OF DIAMONDS one of the most sought after. The Roy Parker artwork carries his trademark touches; the man is winking at the lady. $650.

Midway RACE WAY, October 1963. Midway Manufacturing Company of Franklin Park, Illinois, was formed by former design and engineering employees of United Manufacturing. So they had a running start on the business. Their first pinball, the 2-player RACE WAY, was somehow reminiscent of the Williams HAYBURNERS of over a decade earlier. 3-dimensional toy race cars run around an oval, driven by the players shooting at a line of targets with flippers. $350.

Gottlieb KEWPIE DOLL, October 1960. In with the new and out with the old. Gottlieb tried a single player drum scoring "Score to beat" feature in 1959, and brought it back in 1960. KEWPIE DOLL picked it up, and was also the last Gottlieb game to have the full square backglass and woodrail sides that had appeared in the late '30s. Playfield includes targets, pop bumpers and flippers at the bottom. Artist Roy Parker put the heads of his co-workers on the kewpie dolls. $500.

Gottlieb FLIPPER, October 1960. A game loaded with firsts. It came out the same month as KEWPIE DOLL, yet is radically different. It's the first Gottlieb steelrail, with steel sides (although early production started out with woodrails), the first Add-A-Ball game that gave you additional balls to play for good scores (replacing the replay feature illegal in some areas), the first to have 4 reel drum scoring and the first tapered backbox "Wedge Head" game. $450.

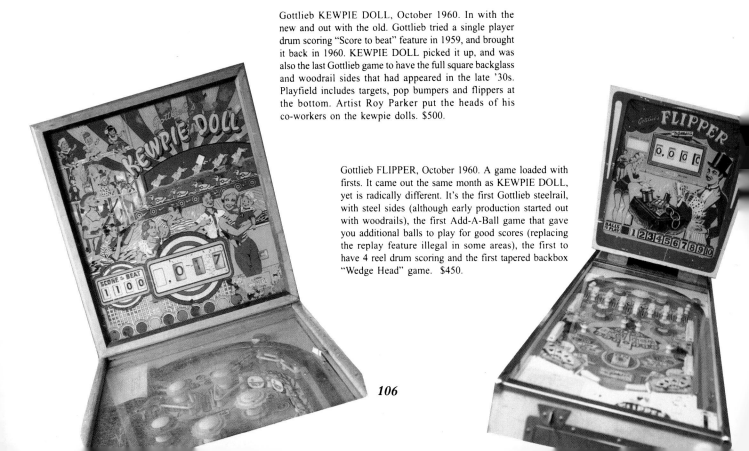

Williams JOLLY ROGER, December 1967. On their way to industry leadership, Williams games often had an edge. JOLLY ROGER features all of the advancements of the '60s, including four player drum scoring, steelrail sides, and the exclusive Williams large, square backglass that offered much more display area than the restrictive Gottlieb "Wedge Head" style. Williams was beginning to hit its stride in pinball design. Available as an Add-A-Ball model. $250.

Williams ACES & KINGS, June 1970. Nobody could match Gottlieb in poker themed pinball machines, although Williams tried. Repeatedly. The oversize backglass gave Williams an advantage, particularly when it came to fitting in 4-digit drum scoring units for four players. Game designer was Steve Kordek, who added a new scoring feature called a "Spinning Post," a center post that scores when rotated. The moody colors don't look much like playing cards. $300.

Midway RODEO, October 1964. Midway pin games evolved into a semblance of conformity with the products of other makers. The 2-player RODEO has drum scoring for each player in a backglass that looks somewhat like a Gottlieb "Wedge Head." But from there the similarities end as the playfield action is entirely different. In the 1980s Midway sold out to Bally. By then their primary business was producing Japanese arcade machines under licence. Their big hit was PAC-MAN. $200.

Williams DOODLE BUG, March 1971. In a paean to the Volkswagon "People's Car," affectionately known throughout the United States as the Doodle Bug (a late honor as Japanese cars were just beginning to arrive to ultimately replace the German car in consumer acceptance), this Williams game continued their spooky colors and hard edge art. Game designer is Norm Clark, who later went to Bally and created the first production solid state electronic pinball game. $300.

Bally DOLLY PARTON, October 1978. After pinball games went electronic, replacing the weight and complications of the electromechanicals with a chip driven solid state system and CRT scoring displays, Bally sought entertainment tie-ins with "Celebrity" games. PLAYBOY and the 6-player SIX MILLION DOLLAR MAN came out in 1978; others (even a STAR TREK) in 1979, including DOLLY PARTON with great art in October. Only DOLLY is a dog to play. Little interest. $200.

Data East MICHAEL JORDAN, September 1992. The most desireable collectible pinballs of the future may well be the short run "Celebrity" games made for special events and people by Data East Pinball in Melrose Park, Illinois, a suburb of Chicago. Five examples of MICHAEL JORDAN were made for use in charity events, and have been sent around the country. They are a modification of the firm's successful LETHAL WEAPON III game being made at the same time. $3,515.

Atari SUPERMAN, April 1979. After PONG, the first popular video game, the Wünderkind of coin-ops was Atari, Inc. of Sunnyvale, California. Pinball games seemed to be the next area of conquest for the firm, so they leapt in with the wide body solid state ATARIANS in March 1977. But they didn't know distribution, and the games had problems. Six games later, with SUPERMAN, they gave up the ghost with as many even newer games still on the drawing boards. $410.

Bally ADAMS FAMILY, October 1992. A movie tie-in pinball, the Bally/Williams ADAMS FAMILY sold more units than any game in the past quarter century. Said to be the best selling pinball game of all time, it isn't. No game will ever match the 50,000 plus units of the Rock-Ola 1933 mechanicals. Still, it is the most successful modern game. Over two years after its introduction it was still rated as the No. 2 game in active commercial play. Collectors are snapping them up. $2,010.

Chicago Coin SOUND STAGE, July 1976. One of the last pinball games made by Chicago Coin before it folded due to competitive pressures. Bell bottoms, rock music and the big backglass styling of Williams couldn't stem the tide. Out they went, while their designers and engineers went to Game Plan, Inc. in Addison, Illinois, and their facilities and inventory to Stern Electronics in Chicago, creating two new pinball makers out of the ashes of one old. $310.

Williams FIREPOWER, March 1980. Steve Ritchie, the designer of the Atari SUPERMAN, went to Williams and created FIREPOWER, bringing a bunch of new ideas with him, including the wide body format. It was the most popular pinball of 1980, and led the way to a creative reorganization of the firm with a staff of young designers. This has led to industry dominance, and the ultimate buyout of its main competitor Bally Pinball. It is said that Williams has an over 80% market share of the pinball sales throughout the world. $460.

Arcade Machines

Arcade machines are a form of coin-operated machine that have long been a thing apart. Their function is to amuse and entertain. They are not gambling machines in the same sense as automatic payout slot machines, or even trade stimulators or counter games with which you can get a multiplied return on your investment if you are lucky. Arcade machines are not luck machines, although they often require a certain degree of skill to play. They generally deliver nothing more or less than what they say they do. You get exactly what you pay for. Most classes of vintage coin machines have a very rigid and limited range of return for your investment, whereas arcade machines can range all over the map in entertainment and amusement while refraining from any of the allurements of other coin machines.

In a way, arcade machines follow a variation of the Sherlock Holmes dictum. Once you eliminate the function and return of the better known coin machines (such as the cash payouts of slots, the cigar or merchandise winning offerings of trade stimulators and counter games, the moving ball of a pinball game, the music of a jukebox, a vended product or service provided by a vending machine, or the revelation of your weight as provided by a coin-op scale) whatever remain must be arcade machines. What remains is a wide open field: games, athletic tests, mechanical automatons, exhibitions of real things or events, shockers, memorabilia stampers, fortune tellers and phrenological head tests, diggers, bowlers and shuffleboards, picture takers and voice recording booths, peep shows and early movies, postcard or picture card and risque card venders, shooting and target and dart throwing and ball bouncing and balloon bursting tests of skill, rides for kiddies, foot vibrators and many other things operated by a coin that haven't been mentioned. It is the broadest class of coin-op, embracing virtually everything that is avoided by the other machine classes. The one common ground is that these coin machines are there to entertain. The Penny Arcades of the past were the places where they were operated, so they picked up the name arcade machines.

The late 19th century examples had no machine class name as the Penny Arcade, really starting around 1905, had not yet been created when the machines first made their appearance. As saloon pieces non-coin-op entertainment and physical testing machines had a long heritage. The "Boys" in the bars loved to prove their superiority to each other, loser buying the drinks. As a result all sorts of bartop athletic and skill tests were developed over the years. Hand grip tests, lung (blowing up a balloon, or weight) tests, devices that measured the strength of a punch and all sorts of other ingenious devices (including Indian Wrestling machines) were common bartop or floor standing pieces. But they didn't bring the house any money; only losers that had to buy. So when the capability of putting coin control into machines came along in the middle 1880s, the saloon games were among the very first to be so fitted. Now whenever a braggart said he had a stronger grip than anyone else, the barkeep pointed to the HAND POWER TEST MACHINE on a stand, and told him to drop in a nickel and prove it. The test made the rounds of the boys in the bar, and the saloon picked up a nickel a pop. The same for peep shows, punch tests (many of which were named after the hero of the age: John L. Sullivan, the boxer), the nippy shockers (which were ultimately banned because they gave some people heart attacks!), lung tests (banned during the TB outbreaks because everyone put their mouth on the same flexible tube lip piece), lifters, bag punchers and all sorts of other athletic test machines.

In the meantime, beyond their early placement in saloons the machines found more and more usage in amusement centers, often located on the edge of town at the end of the streetcar line where the transit company built an amusement park to get a double fare from the riders, both going and coming. The machines they used started out with early amusements, and depictions of the new coming age of technological wonders. By the early 1900s the amusement parks were picking up nickels by the drove with their peep shows and MUTOSCOPES, hauling in "white money" (nickel, dime and quarter coins) as well on machines that allowed you to play a soccer game with a partner, or manipulate boxers, or even (a very early game, quickly removed) have a two-sided COCK FIGHT.

Dennison MUSICAL FAIRY, 1885. Exhibition coin machines, the earliest form of arcade machines, started in England in the late 19th century. A major maker was John Dennison of Banks, Leeds, England, producing machines between 1871 and the turn of the century. Clockwork mechanism gearboxes move the figures once a coin is dropped. This detail of MUSICAL FAIRY shows the doll that raises her hand, at which point shutters to the left behind the fairy open to a reveal a fortune. $4,510.

Exhibition machines included a fully articulated locomotive, first placed in railroad stations in the middle 1880s, with a steamboat version in one of the buildings at Coney Island. A Waukegan, Illinois machine maker named Edward Amet produced simulations of a yacht race, and a miniature power station that lit up electric "Lightoliers" once the station got going after two pennies were dropped. These marvelous machines soon had their own dedicated parlors in major cities, to be followed in small towns, with the Penny Arcades providing a complete array of "Automatic Vaudeville" for the masses.

The leading producer at the time was American Mutoscope & Biograph Company, making the MUTOSCOPE peep show. The basic mechanism was developed in the late 1890s, and modified over the years. The circa 1905 original machines were cast iron, but broke easily. When the International Mutoscope Reel Company Inc. was reformed under new ownership in the early 1920s a steel cabinet replacement was created to keep the machines running. The ALL STEEL MUTOSCOPE was first made in 1926 and continued well into the 1930s. It was nicknamed the "Tin (only it wasn't) Mutoscope." In the United States these machines are called "Peep Shows," but in England, where they sold almost as many MUTOSCOPE machines as they did in America, they are called "What The Butler Saw Machines." Originally the picture fare was characteristic of the times, showing world's fairs, amusement parks, lovely showgirls and domestic situations. Early on it was for families, but increasingly the peep shows and MUTOSCOPE reels got a bit racy, until the arcades had a bad name. By World War I they had started to degenerate to side street entertainments on the sleazy side, and remained so to some degree through the '20s and '30s. Competition was keen, with many low cost competitors to the American Mutoscope MUTOSCOPE.

One such producer was the American Novelty Company of Cincinnati, Ohio. American made about 7 or 8 varieties of peep shows, all under the same 1913 patent. The earlier version was called the SCULPTOSCOPE and was a nickel-plated model that stood high. It was made between 1908 and 1924. The later model was called the WHITING VIEWER and was much smaller, painted black. Due to its odd shape it was nicknamed the "Acorn." It was made from 1922 up to about 1939 when American came out with an even sleeker version.

Popular culture was also an arcade machine driver, with striker games similar to the old carnival hammer games that lifted a weight to ring a bell becoming classic arcade pieces. One variation was made by B. Madorsky of Brooklyn who produced a game called the LINDY STRIKER in 1929, named after the American aviation hero Charles A. Lindbergh, who in 1927 was the first pilot to fly solo across the Atlantic Ocean. Lindbergh popularized flying, and Madorsky among others cashed in on his fame.

After that came one of the most significant developments in arcade machines, the digger. The digger was a large glass-walled boxed machine that held all sorts of toys and personal items, such as fountain pens, pocket knives and cigarette lighters. All you had to do was drop your nickel, turn the crank and direct the "claw" to the item, and maybe win it. Except the claw never seemed to follow your control, and winning was difficult. Some states even outlawed the machines as gambling devices. But by-and-large they became the primary arcade piece in the '30s and into the World War II years.

Other shoot-em-up arcade machines regained the macho gun and peep show image of the past during the World War II years, and for a few years afterward. By the '50s and '60s arcade attention had gone downhill, both because of game boredom and the arrival of television and home entertainment. It wasn't until the advent of the video game in the 1970s, and the fast and furious interactive Atari PONG followed by the largely Japanese machines of the '80s and '90s, that the arcades once again gained favor and began to move back into the mainstream of American entertainment. There was one year back there, at the end of the '70s, where arcade machines brought in more money than the movie industry.

Dennison SORCERER FORTUNE TELLER, 1887. The weight of a coin dropped on the lever at right releases the clockwork mechanism, left, shown with the cabinet front removed. Once the action has started the SORCERER drops a printed fortune down the center chute to the player. Few of these machines have made it to the Americas, although they are highly sought after and major collectibles in the U. K. and former Commonwealth nations. Valuable machines. $4,510.

The upcoming popularization of virtual reality machines (for which we will pay $2 to $5 per game, or even a lot more) promises to provide a true rebirth of arcade machines in shopping mall locations, with Japan once again pointing the way. A recent survey conducted in 1992 revealed that 79% of teenage girls in Japan visited an arcade at some time during the year. Hang onto your hat. This characteristic is coming to the rest of the world.

Arcade machine collectibles go back as far as the 1880s, and reach right up into the 1980s and beyond. This is one of the few areas of technical collectibility (and certainly coin-op) that is really over a hundred years old. The amazing fact is that most of the vintage arcade machine finds of the present are of machines from the 1905 through 1940 period that were stuck in barns, attics, basements, warehouses and other locations in the past and left there to rot. After years of neglect, with only a few dedicated collectors who loved the machines, arcade machines are now coming to the fore with a rush to become one of the hottest coin-op collectible areas of the future. Even some of the early video games are becoming valued collectibles, certainly the early chip driven games of the 1970s. While most of them aren't worth the money to haul them to the dump, certain games are beginning to gain stature because they are fun to play. Now is the time to find them and keep them, because they will most certainly be worth a lot more in the future.

Farrington GRIP TEST, 1886. American coin-op amusement machines, leading to the arcade class, started out as coin controlled versions of saloon and cigar counter entertainments, enabling the operator to get paid for the physical strength tests so enjoyed by their clientele. The Farrington Cigar Company combination GRIP TEST and cigar clipper, patented on June 22, 1886, was profusely copied and made in coin-op versions on both sides of the Atlantic. $2,210.

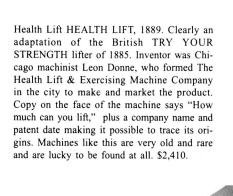

Health Lift HEALTH LIFT, 1889. Clearly an adaptation of the British TRY YOUR STRENGTH lifter of 1885. Inventor was Chicago machinist Leon Donne, who formed The Health Lift & Exercising Machine Company in the city to make and market the product. Copy on the face of the machine says "How much can you lift," plus a company name and patent date making it possible to trace its origins. Machines like this are very old and rare and are lucky to be found at all. $2,410.

Rowland TRY YOUR STRENGTH, 1885. An invention of Londoner Robert E. Page, TRY YOUR STRENGTH was produced by A. Rowland & Son, London, England in the late 1880s. It established a format for lifting machines that endured for a century, and was copied on both sides of the Atlantic. The lettering "TRY your STRENGTH" is in the casting supporting the dial box, turned on its side. The same mechanism reversed, to push instead of pull, was used as a coin-op scale. $4,015.

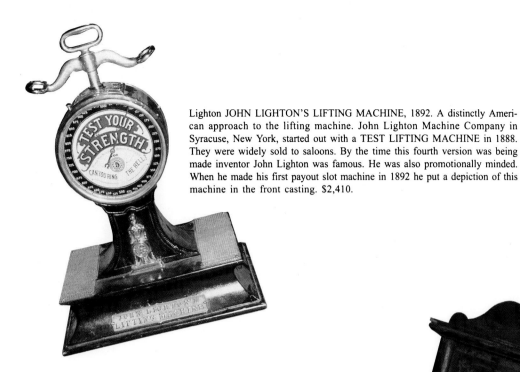

Lighton JOHN LIGHTON'S LIFTING MACHINE, 1892. A distinctly American approach to the lifting machine. John Lighton Machine Company in Syracuse, New York, started out with a TEST LIFTING MACHINE in 1888. They were widely sold to saloons. By the time this fourth version was being made inventor John Lighton was famous. He was also promotionally minded. When he made his first payout slot machine in 1892 he put a depiction of this machine in the front casting. $2,410.

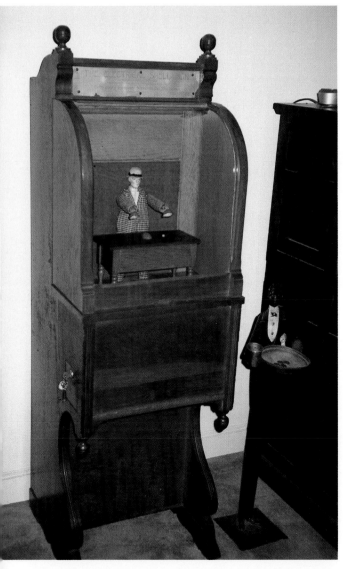

Colby COMBINATION LUNG TESTER, 1892. Prolific early inventor of large and small coin controlled machines, John Colby was associated with a confusing array of manufacturing and operating companies based on the origin of the seed money. Colby Specialty Supply Company of Chicago, Illinois, made one of his most popular machines, a combination lung test and dice thrower for the saloon crowd. A cache of these machines was found in an old upstate Michigan hotel in the late 1980s. $4,510.

National Automaton THE ELECTRIC SHELL MAN, 1893. The grifters that travelled the towns out west were experts at the old three shell game. It became the expected barroom scam. The game was mechanized and electrified by The National Automaton Company of Washington C. H. (meaning Court House), Ohio. Much like a 3-reel slot machine, drums and pulleys made the pea under the shell appear to move mysteriously about. Bar patrons bet on the outcome, which changed. $6,520.

Auto-Machine AUTO-MACHINE, 1897. The idea of seeing real pictures of racy ladies for a nickel was enough to make a whole raft of peep shows successful. AUTO-MACHINE was made by the American Auto-Machine Company of New York City, and was sold through the mail to saloons all over the country. You dropped a nickel in the coin slot at the top, and stuck your eye in the peep hole to see the show. Peep shows pre-dated the movies and paved the way for visual entertainment. $7,520.

Auto-Machine AUTO-MACHINE, 1897. The workings of the AUTO-MACHINE are very simple. The nickel trips a lever, which activates the battery driven light and releases a clockwork mechanism that gives you a few seconds look at 15 different glass slides showing women in tights and striped stockings, one even smoking. Wild stuff! The only known example of this machine was found in Madison, Wisconsin, in the early 1990s and quickly made it to an advanced collector.

Roovers MADAME ZITA, 1895. Roovers Brothers of Brooklyn, New York, made strength testers, shockers and scales. They ultimately settled into the business of making coin-op automaton fortune tellers. MADAME ZITA was a purely mechanical machine, operating by a handle pull. In later years it was electrified to move the figure and dispense the printed fortune ticket, at which point the original model became known as the "Mechanical Zita." Rare and valuable. $18,040.

Watson MEDICAL BATTERY, 1897. In the early days of electricity, after lighting entered the home and streetcars provided the first cheap public transportation, wonderful medical powers were attributed to the power source. The result was electric treatment machines, known as shockers, as they gave you a jolt of voltage for a coin. Watson Machine & Novelty Works of Detroit, Michigan, made the MEDICAL BATTERY and sold them to drug stores. Most of them have been found there. $1,210.

Sapho SAPHO, 1898. First of the large commercial floor standing peep shows, SAPHO displays four sets of 12 stereo cards which you can see for a penny a set. Maker was the Sapho Manufacturing Company of Chicago, Illinois, a small independent firm. The success of the machine was so great it was quickly licensed to a number of other Chicago manufacturers, including the Mills Novelty Company, Paul E. Berger Manufacturing Company, R. J. White and others.

Sapho SAPHO, 1898. One of the secrets of the SAPHO's success was its come-on. The four shows appeared in sequence, next one up. After each show a potential viewer could press a buzzer button on top and see the first stereo of the next program free, but it took a penny to see the rest of it. So the first card was always great. SAPHO introduced fine cabinet work and elaborate castings to the peep show. Few surviving machines are known. $3,410.

Berger THE MAGIC FORTUNE TELLER, 1899. Paul E. Berger Manufacturing Company of Chicago was the largest electric floor model gambling machine maker. By 1899 they were rapidly being replaced by the Mills OWL or other machines. Using the CHICAGO RIDGE model, Berger put answers to questions on the dial and provided a mask so only one could be seen at a time to convert the payout to a fortune teller. Ten questions on the front, such as "Will I be rich?" were answered, in this case by "Not in this country." $5,510.

Kansas City PEEPOSCOPE, 1899. Peep show interest and production capability was rapidly dispersed. The Kansas City Talking Machine Company of Kansas City, Missouri, was an Edison Talking Machine distributor that went off on their own to create early movies and peep show programs, making PEEPOSCOPE bartop viewing machines to promote their products. The large cattle driving and railroad labor forces from Sedalia, Missouri to Kansas City and west provided a ready saloon market. $910.

Midland ELECTRICITY IS LIFE, 1899. Most popular early shocker was the Midland ELECTRICITY IS LIFE, a reference to the believed powers of electricity as a cure all. Midland Manufacturing Company was in Chicago, Illinois, and made these battery operated shock machines for saloons, drug stores, cigar counters, restaurants and the like. Drop a penny in the top, grab both knobs, push the right one down to get current and stop when you can't take more. $1,010.

American Mutoscope MUTOSCOPE MODEL A, 1899. The Crown Prince of peep shows, and the one that survived the longest, was the MUTOSCOPE, a flip card crank driven action "movie" machine that spun photographic images on a reel of cards attached to a wooden spool fast enough to suggest actual motion. It was the wonder of its age. Maker was American Mutoscope Company of New York, New York. The first three A models were in wooden cabinets, known as the "Oak Mutoscope." $1,210.

Midland NEW CENTURY STRENGTH TESTING MACHINE, 1900. It takes up to four coins to get everything out of (or put everything into) this device. The four time combination includes two arm tests and two grip tests, with the the results showing up on the dial at the top, highlighted by small bulbs that light up as the pointer moves higher up the dial. These machines were virtually unknown to collectors until the first one showed up in the mid-1980s. $1,210.

New England TEST YOUR STRENGTH, 1900. Small enough to be a saloon game and not get in the way of pouring beer, yet strong enough to try the strength of the toughest. A popular bartop of the early 1900s. Drop a penny, grip the two knobs on the sides, and twist. It's enough to make your eyes pop! But if you falter, or try and get a better grip, the pointer locks on the dial and gives the next guy something to beat. $3,410.

Mills Novelty NEW OWL LIFTER, 1901. Primarily a gambling machine maker, with their OWL floor machine becoming the best selling slot machine of 1898, Mills Novelty Company of Chicago made the owl symbol their logotype. Their first arcade model was made in 1899 and named after the slot machine, becoming the OWL LIFTER. Improved in 1900, it was redesigned in 1901 as the NEW OWL LIFTER to incorporate a marvelous casting of an owl's head over a new dial. Elegant and elaborate. $3,610.

American Mutoscope MUTOSCOPE MODEL B, 1901. The most successful arcade machine ever made, with many still in use in theme parks. MODEL B has a cast iron cabinet with its coin chute on the left side, called the "Clamshell" based on its design. MODEL DL added improvements, and a base called the "Iron Horse." TYPE E has an "Eagle" base and improved lighting, with its coin chute moved to the right side. Highly desirable, particularly with original picture reels. $2,510.

Peterson MEDICAL ELECTRO BATTERY, 1904. The level of electricity increased when you held the grips of the MEDICAL ELECTRO BATTERY once the "treatment" start lever was pushed. When you finally released (the sign on the dial says "How much can you stand?") the pointer stopped in place for all the world to see. Maker was T. W. Peterson Company of Chicago. The prairie city was loaded with small makers of machines who learned their craft at Mills Novelty or other Chicago firms. $1,810.

Mills Novelty IMPERIAL, 1902. There was always a market for a shocker, so Mills Novelty bought the rights to the Midland ELECTRICITY IS LIFE and made their own version in a very attractive cast iron cabinet instead of wood. It came out in 1900 with a scene of the city of Chicago as a photographic background. When they shifted the background scene to a woman, it sold better. That makes the "Cities" IMPERIAL rarer than the "Girl" machine of two years later.

Roovers PUSS-IN-BOOTS, 1904. A character right out of a favored child's story, "Puss" is a fortune teller that moves, nods its head, turns to a stacked column to pick up and drop a printed reading into the cup. Very animated, and a fabulously popular machine in its day. They are rare and valuable, as are all Roovers machines. Their market was international, sold by Roovers and Mills Novelty. This example was found in Australia. $18,040.

Caille Bros. CAIL-O-SCOPE, 1904. After the great success of the SAPHO and MUTOSCOPE machines all of the major makers created their own floor model peep shows. The Caille Brothers Company of Detroit was second only to Mills Novelty (but not a close second) as the largest arcade machine producer. CAIL-O-SCOPE offered a 15 view show for the money, flopping cards in sequence. "What he saw through the keyhole" was a typical presentation. And he didn't see much. $2,810.

Marvin & Casler NEW PALM READER, 1905. American Mutoscope didn't make their MUTOSCOPE machines. Marvin & Casler Company in upstate Canastota, New York, and other subcontractors, did. The M & C firm did a lot of the engineering. They also made an incredible and mysterious all mechanical machine called PALM READER, produced in a much larger floor model as NEW PALM READER. Place your hand palm down and small steel rods lift up to record its contours and give you a reading. $6,215.

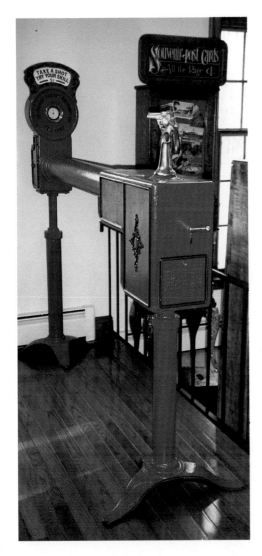

Mills LION HEAD LUNG TESTER, 1904. Saloon placements were diminishing and dedicated Penny Arcade locations were on the rise by 1904. It changed the machines. Mills Novelty had become the largest producer, with exotic models made by no one else. The massive and very commanding LION HEAD LUNG TESTER is an example. Pay a penny, blow in the tube, and see how long you can blow and drive the pointer. If you hang in there the lion roars and his eyes light up. $65,040.

Automatic Target AUTOMATIC PISTOL, 1905. It took an arcade to operate this piece as it takes up more room than a saloon could afford. The Automatic Target Machine Company of New York City made their first gun games in 1892. AUTOMATIC PISTOL has a dozen years of development behind it. Drop the coin, push the re-set lever and you get a shot. Bang! It's loud. And a bullet hole shows up on the target. How did they do that? Depending on your aim, "Hits" are flipped out. $11,030.

Mills WIZARD, 1904. WIZARD promised to "answer questions intelligently" after the pap answers of the past, going beyond "Yes" or "No" or "Not in this country." Drop a penny in the slot, turn the pointer to your question, push the plunger at the right, and it spins out an answer. Question and answer inserts were periodically changed to keep things fresh. WIZARD has started a number of arcade machine collections as they are fairly common and relatively inexpensive. $510.

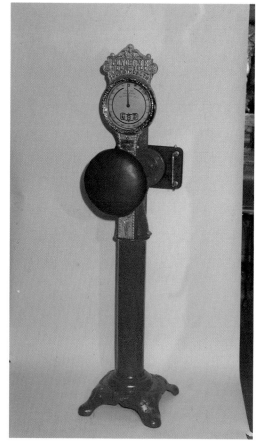

Mills Novelty 1905 AUTOSTEREOSCOPE, 1905. The spring wound AUTOSTEREOSCOPE was first made in 1904, showing only one set of cards rather than the four of the SAPHO and similar Mills QUARTOSCOPE. The machine was electrified as the 1905 AUTOSTEREOSCOPE and went on to a long and useful career. Machines like this operated well into the World War II years, kept up to date with newer card sets. Most of those found come from old amusement parks that were still running them. $1,410.

Caille Bros. OLYMPIA PUNCHER, 1906. One of the favorites of the barroom crowd, and an acceptable way of testing a man's punch without sticking your jaw in the way. It was something to bet on. So the players gave it their all. Sometimes, when they missed, they broke their hand on the dial or the wall. The original OLYMPIC PUNCHER of 1904 didn't have hand protection. They added hand grips to create the OLYMPIA PUNCHER so you wouldn't spin if you missed. $5,515.

Caille Bros. WHEEL OF FATE "Gentlemen", 1907. The major slot machine makers all converted gambling machines to fortune tellers. The randomness of the spin fit the needs of fortune telling well as nothing was preprogrammed. In all likelihood Caille Bros. converted their VICTOR "Slant Front" floor slot to fortunes because it wasn't selling well as a payout slot, so they had the parts around. A "Ladies" version and MUSICAL WHEEL OF FATE models were also made. $5,515.

Caille Bros. COUNTER COMBINATION BLOW AND GRIP MACHINE, 1905. Arcade machines often have long names designed to explain what they do, and the more they do the longer the name. The floor model of this dual action grip and lung test is ELECTRIC COMBINATION BLOW AND GRIP. A tower of electric lights was added to the model, called MASCOT ELECTRIC TOWER GRIP. The "Mascot" name is quicker, so that's what collectors call them. Colored lights flash on as the dial advances. $3,210.

Mills Novelty NEW GRIP MACHINE, 1906. The Mills Novelty answer to the Caille counter "Mascot" flashes lights around a circle as long as the grip is held, requiring tightening as time goes by. The dial says "Test your grip. How many lamps can you light up." Such innocent and demanding pubic amusements would hardly get the time of day today, but when electricity was young and machines were marvels things like this really caught on. One bulb lights all the colors. $2,210.

Caille Bros. MYSTIC WHEELS (GENTLEMEN!), 1906. To take advantage of random reel spinning Caille took their popular ROYAL JUMBO card machine and converted it to a three reel fortune teller; two wide with fortunes and a narrow center reel with photographs. The whole idea is to give the player some readings plus a peek at their future mate in life. That's assuming they were all single. The GENTLEMEN! machine showed pictures of ladies, and the LADIES! machine showed men. $4,510.

Hance TARGET PRACTICE, 1923. Starting in the 1890s counter gun games became a major arcade format in the 1920s with a dozen makers. Hance Manufacturing Company of Westerville, Ohio, was a glass globe peanut and gumball vending machine maker that produced gun games and private label venders on the side. TARGET PRACTICE is notable for its all aluminum cabinet. The more desirable PENNY-BACK TARGET PRACTICE returns your money in a cup on the side for a good score. $1,210.

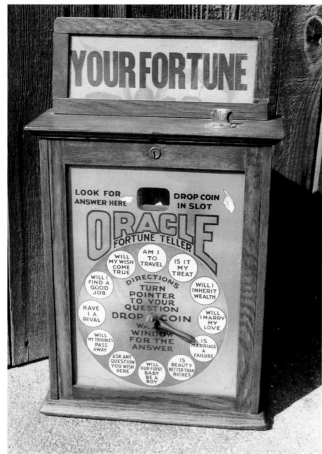

Exhibit Supply ORACLE, 1922. New arcade machine development all but halted with World War I. When the war was over both Mills Novelty and Caille Bros. reduced their commitment to the machines, soon dropping them. Starting in 1917, the Chicago based Exhibit Supply Company stepped up to become the leading supplier, a position it held until World War II. ORACLE is a countertop fortune teller used in arcades and travelling carnivals. $410.

Gatter 1926 TEN PIN GUM VENDER, 1926. Rudolph Gatter was a big name in arcade machines, designing many for himself and other firms, including Exhibit Supply Company as well as International Mutoscope. Early on he had his own firms. He made his first AUTOMATIC BOWLING ALLEY in 1902. The much improved 1926 TEN PIN GUM VENDER was made by Gatter Novelty Company in New York City. The large display area at the back was filled with gumballs, dispensing one with each play. $510.

Harvard METAL STAMPER, 1923. A takeaway souvenir from a Penny Arcade visit was a popular idea, and the spell name tokens often show up at antique estate sales. The machines were first made in the early 1900s by the Heene Automatic Stamping Machine Company, Phoenix, Arizona. The firm moved to Cleveland, on Harvard Street, and became The Harvard Automatic Machine Company. The filigree legs on this later model HARVARD STAMPER gave it the collector nickname "Ornate." $2,010.

Chester-Pollard PLAY FOOTBALL, 1926. Close-up of the playing figures in the Chester-Pollard PLAY FOOTBALL reveals their soccer garb. This is a very competitive and exciting 2-player game, which is probably why it has survived the years so well. Very desirable collectible. You push the handle down for your kickers to kick the steel ball, and your opponent does the same with the other handle. Three scores, with the odd to avoid ties, ends the game.

Chester-Pollard PLAY FOOTBALL, 1926. One of the most popular arcade games of all time, with one still operated at Disneyworld in the late 1980s. It looks British, particularly since the football game of the title is actually a game of soccer (called football in the U.K., Europe and elsewhere outside of the United States). But its patents are American, with production by the Chester-Pollard Amusement Company, Inc. of New York, New York. $2,210.

Mills Novelty BAG PUNCHER, 1926. Originally made for saloons in 1902 as the PUNCHING MACHINE, including a payout model, the Mills coin-op punching bag was upgraded for Penny Arcade placement. The BAG PUNCHER of 1926 had a completely new cabinet and attractive display graphics more fitting a public entertainment area. Boxing was popular in the '20s, with the broadcast of the Dempsey-Tunney fight on September 23, 1926 pulling the largest radio audience to date. $2,710.

International Mutoscope ALL STEEL MUTOSCOPE, 1926. American Mutoscope And Biograph Company got out of the MUTOSCOPE business in 1909 to become the American Biograph Company, a motion picture pioneer. The machine business was finally sold in 1923, and the International Mutoscope Reel Company was formed to service the surviving arcades. The machines were in such poor shape the components were used to create new models in a steel cabinet. British arcades sometimes painted the sides of their newer machines in the racy style of the reels. This is folk art. $1,210.

Doraldina PRINCESS DORALDINA, 1928. The ultimate development of the Lady-In-A-Box fortune tellers started by the Roovers MADAME ZITA in 1895. What makes PRINCESS DORALDINA different is that you can feel close to this woman. She is life size, attractive, and makes human moves as she scans the cards for a peek into your future, and then your fortune comes out of the cup on the front so that you have something to take away for the experience.

Doraldina PRINCESS DORALDINA, 1928. One of the most intriguing features of PRINCESS DORALDINA is the fact that she appears to be breathing. Her chest heaves as she moves, and her eyes roll. Maker was the Doraldina Corporation in Rochester, New York. The "Princess Doraldina's Prophecies" dropped in the cup are 4-page folders that many people kept as true readings. They are sometimes found among miscellaneous papers at farm sales and estate auctions. $12,030.

Edwards GRIP TESTER, 1929. Prohibition took away America's saloons in 1920, but not the old games. Only now they weren't a test of strength that made the loser buy a round of drinks; they were health machines. Of sorts. The grip test came back in the '20s in a number of guises with more makers than it had in the earlier drinking days. Among them was the Edwards Novelty Company of Fort Worth, Texas, with side copy that suggested you should "Test your grip every day." $260.

Norris MASTER TARGET PRACTICE, 1928. The Norris Manufacturing Company of Columbus, Ohio, was noted for their MASTER counter peanut and gumball vending machines, starting around 1922. Norris TARGET PRACTICE gun games were a parallel line, with various models from 1923 to World War II. This battery operated penny back version has lights that indicate the score when hits are made. The penny is returned in the cup at the back. Hance was a Norris private label producer. $810.

Chester-Pollard PLAY DERBY, 1929. Following the success of the PLAY FOOTBALL in 1926, Chester-Pollard was off and running as a major arcade machine producer. PLAY DERBY was one of a series of large cabinet games based on sports and sporting events. It is a 2-player competitive game, with each player winding a crank to move their horse along an oval track to the finish line. Great stuff for an outing in the '20s as there was virtually no home entertainment except radio. $3,815.

Peo VEST POCKET BASKETBALL, 1929. Radio made the whole nation sports minded as it provided a feeling of real time involvement. In the search for broadcast sports basketball was suddenly discovered. Peo Manufacturing Corporation in Rochester, New York, made a series of countertop games based on various sports. VEST POCKET BASKETBALL was one of its most popular, and won a silver cup at the 1929 convention of coin machine operators. These sometimes show up at yard sales. $460.

Peo BARN YARD GOLF, 1931. When Peo Manufacturing Corporation introduced a new counter game called LITTLE WHIRLWIND in 1930 it was revolutionary. Using the old whirlwind and PARLOR KEEPS idea, Peo made the playfield almost vertical, adding great speed to the game. It was produced in a wide variety of models. Just as others copied it BARN YARD GOLF changed the format to drop the ball in the middle, scoring in troughs to fill out a good poker hand. $310.

Atlas BASE BALL, 1931. With radio, the most popular sport of all was baseball, the "All American Game." Atlas Indicator Works of Chicago made a killing on a large 1930 floor model arcade baseball game called BASEBALL. So they came back with a small penny flip countertop that used the coin for scoring. Put in a penny, pull the lever and release and the coin jumps onto the playfield diamond. Bead counters at the bottom kept track of the score. $560.

Pace DUCE WILD, 1931. Proving you couldn't keep a good idea to yourself, patented or not, the Peo BARN YARD GOLF idea was just as widely copied as the earlier LITTLE WHIRLWIND. Pace Manufacturing Company in Chicago, primarily a slot machine and coin-op scale maker, reproduced them all. LITTLE WHIRLWIND was made as WHIZ BALL and BARN YARD GOLF was produced as DUCE WILD. These whirlwind games were sold in great numbers and are an entry level vintage arcade collectible. $360.

Exhibit Supply 20TH CENTURY, 1933. The 1920s made Exhibit Supply the arcade machine leader, and a pioneer in the prize grabbing digger, producing a series of IRON CLAW models. Seeking to expand the market to restaurants, stores and hotel lobbies they produced a smaller, less expensive 20TH CENTURY model in 1933 in a beautiful quarter sawed oak cabinet. Drop a penny, guide the claw with the dial, then it twitches and drops the prize before it drops it in the chute. Typical. $2,760.

International Mutoscope 1934 CHAMPION, 1934. Running a close second to Exhibit Supply as the leader in arcade machines, the International Mutoscope And Reel Company practically paralleled their lines. Their economy digger model was the 1934 CHAMPION, with a colorful Deco design prize chute. A clever addition is the Automatic Candy Vender to the right of the prize chute and right below the control wheel. It made the machine a vender and not a gambling device.

Exhibit Supply ELECTRIC EYE, 1936. Just skirting the edge of a gambling machine, the greatly enlarged ELECTRIC EYE counter gun game shoots beams of light rather than coins or pellets. As a new technology it was as interesting as the video games were some 35 years later. Looking like an overgrown A. B. T. gun game under glass, the colorful field has dial pointer counters for shots expended and hits made. The player shot for score. $660.

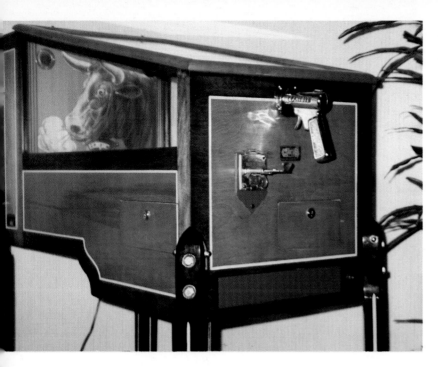

Exhibit Supply ELECTRIC EYE, 1936. Target for the ELECTRIC EYE is this snorting head, which means you are always shooting the bull. It is difficult to aim as you have no idea where the bolt of light is going. Later models had big signs over them that explained the game. An automatic payout model that returned 2-4-6-10-20 coins in a cup was produced as JACKPOT ELECTRIC EYE. It is worth considerably more to collectors. $660.

Exhibit Supply CRYSTAL PALACE, 1939. This is the highest visibility digger machine ever made. The idea was to show off the contents inside to attract players. So Exhibit Supply created a full top glass cabinet that allowed you to look inside from all four sides to pick out that special prize you wanted. Diggers are highly desirable collectibles as they not only look good but provide a lot of entertainment. Collectors like to fill them with 1930s items. $1,815.

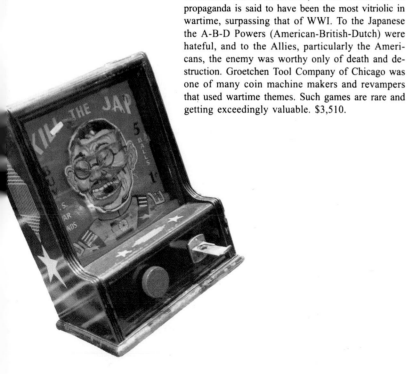

Groetchen KILL THE JAP, 1943. World War II propaganda is said to have been the most vitriolic in wartime, surpassing that of WWI. To the Japanese the A-B-D Powers (American-British-Dutch) were hateful, and to the Allies, particularly the Americans, the enemy was worthy only of death and destruction. Groetchen Tool Company of Chicago was one of many coin machine makers and revampers that used wartime themes. Such games are rare and getting exceedingly valuable. $3,510.

Seeburg CHICKEN SAM, 1939. Gun games got bigger and bigger just as the world was on the verge of war. The largest was the CHICKEN SAM made by J. P. Seeburg Corporation in Chicago, a jukebox producer trying to get into other fields. This electric eye game has a hobo named Sam stealing chickens. Hit his target and he spins around and goes back where he came from, back and forth. Before plastic, the figure is of carved wood. Wartime revamps converted him to Hitler or Tojo. $710.

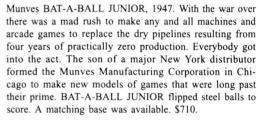

Exhibit Supply MONARCH MERCHANDISER, 1940. The last digger model produced by Exhibit supply before wartime material restrictions ended coin machine production early in 1942. It became the generic cabinet for upgraded and repaired diggers during the war years, so it insides are often those of a variety of models. The inside crane isn't as detailed as pre-war models, and the trim doesn't include a lot of castings, probably to save critical materials. $2,210.

Mann SUPER GRIP, 1947. Even the venerable strength test machines with a heritage going back to pre-Prohibition saloon days went into production right after the war. The Mann Novelty Company located on the south side of Chicago made something they called an "Arcade Strength Test Scale" called SUPER GRIP, in effect the old grip test machine. A stand alone streamlined matching base was also available. Machines like these have little value but are great for a bar. $110.

Munves BAT-A-BALL JUNIOR, 1947. With the war over there was a mad rush to make any and all machines and arcade games to replace the dry pipelines resulting from four years of practically zero production. Everybody got into the act. The son of a major New York distributor formed the Munves Manufacturing Corporation in Chicago to make new models of games that were long past their prime. BAT-A-BALL JUNIOR flipped steel balls to score. A matching base was available. $710.

International Mutoscope DROP KICK, 1949. Wanna be a football hero? Just slide a nickel in this thing, and when the ball is released give it a good swift kick. But don't miss or you'll end up with a mangled foot. Maybe that's why this machine is so rare. The wording on the front says "How far can you kick." It was measured by impact, much as the old saloon punchers were years before. Kick it to hit the 500,000 mark at the top of the scale and you're an All American. $1,210.

Marvel SLUGGER, 1948. The Marvel Manufacturing Company was formed in Chicago in the war years to revamp pinball games to give them new names, playfields and backglasses. After the war they embarked on their own games, picking up public domain patented ideas from years back. They made a series of clever counter pinfield baseball games called POP UP, SLUGGER and a few others. The dials hit the ball and moved the catcher across the playing field to make the catch. $460.

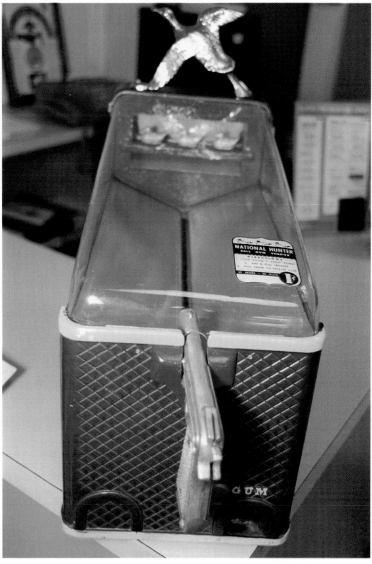

Silver King HUNTER, 1949. New materials and makers posed a postwar threat to existing long term producers. A classic example is the redesign of the A. B. T. counter gun format in a plastic cabinet. Silver King Corporation of Chicago was a postwar maker of gumball machines, a light-up grip test (that flashed pictures of Burlesque Queens!) and the redesigned HUNTER gun. A duck casting was at the far end to make the point. A gumball was dispensed with every play. Common. $160.

United 10TH INNING, 1949. United Manufacturing Company was also formed during World War II to revamp pinball games. Partner Harry E. Williams was particularly good at baseball games. Partner Lyn Durant did the circuitry and production engineering. Then Williams went off on his own, and Durant turned out his first baseball game in 1949. He didn't have the touch and it looks like something out of the mid-'30s. But the circuitry, ball kickers and play are great. $460.

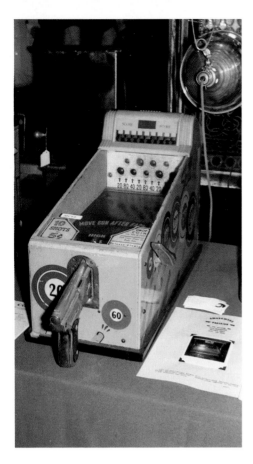

A. B. T. ELECTRIC SKILL GUN, 1949. A. B. T. Manufacturing Corporation was located in Chicago. Best known of the counter gun makers, with models continually evolving from the mid-1920s through the '60s, the post-war A. B. T. models were improvements over the pre-war guns of the early '40s. SKILL GUN (the "ELECTRIC" part of the name was rarely used) totals the score and lights up the hits. It was sold to school stores, arcades and factory rec rooms. $410.

Genco BOWLING LEAGUE, 1949. The tavern became a mixed company entertainment center after WWII, and participating games were all the rage. Bowlers became hot. Genco Inc. had been formed at the beginning of the pinball boom in 1931 to make copycat games at lower cost. Their philosophy held true after WWII. With everyone making shuffle bowlers, Genco entered the arena with BOWLING LEAGUE featuring automatic backglass scoring and puck return. $265.

Auto-Photo AUTO-PHOTO NO. 7, 1950. Taking your own photograph in a coin-op goes back to the early 1900s. It was popularized in the late 1930s by the International Mutoscope PHOTOMATIC, a self-contained booth placed in railroad stations and bus depots during the WWII years. Many a serviceman and woman sent pictures home in PHOTOMATIC aluminum frames. The Auto-Photo Company of Los Angeles carried on the tradition. These machines need a lot of room. $920.

United IMPERIAL SHUFFLE ALLEY, 1953. The first big postwar boom in arcade machines was with shuffles, a tavern necessity in the '50s. The biggest producer was United Manufacturing in Chicago, starting with their SHUFFLE ALLEY in 1949. For the next ten years United made dozens of newer versions, all somehow including SHUFFLE ALLEY in the name, led by DELUXE, TEAM, ROYAL and other prefixes. IMPERIAL SHUFFLE ALLEY is a 6-player with a triple match feature. $260.

Exhibit Supply BIG BRONCHO, 1951. Postwar shopping dispersion, moved out of city centers to the sprawl of suburban shopping malls reached by automobile, created new locations for amusement machines. The "Kiddie Ride" gave shopping mothers a brief break while the kids rocked. One of the few mechanical arcade machine types that have survived the video era. Made by Exhibit Supply, Bally, Bert Lane and others. These make great rocking horses for grandchildren. $380.

Williams JOLLY JOKER, 1955. A machine class with a very short life. "Rolldowns" give the player a large rubber ball in the fashion of an old carnival game, which the player then rolls up the glass to drop down at the top and roll back down, making scoring holes as it traverses the course. JOLLY JOKER lit up a poker hand as holes were made. The problem was people stole the balls. Williams Manufacturing Company of Chicago used the name again in a 1962 pinball game. $260.

Bally SUPER BOWLER, 1958. While most bowlers of the day were uncovered to give the players the feel of the ball and a measure of control, Bally put the game under glass to curtail ball stealing. The reduced size SUPER BOWLER came in two models with the same basic backglass for limited space placement. ALL STAR DELUXE BOWLER shoots the ball with a fast reloading "ball gun" and scores like a bowling game. SUPER BOWLER adds a match-score feature. $410.

United CARNIVAL GUN, 1954. Only a few gun games were available during the WWII years. But after the war, and the Korean War, they proliferated. One reason was the increased technology that let the shooter "see" the targets at a distance in three dimensions, whereas they were actually folded mirrored optics. CARNIVAL GUN gives the player the shots offered at a traditional carnival shooting range, including ducks, squirrels, rabbits and a bullseye. $310.

Chicago Coin CARNIVAL RIFLE, 1968. By and large Chicago Coin just copied. It did at least provide a better indication of what was selling than any industry report. CARNIVAL RIFLE is a takeoff of the earlier United CARNIVAL GUN with drum reel scoring replacing the lighted backglass scoring of the 1954 game. A new feature was the "Pulsating Windmill-Bullseye" that boosted scores and awarded extra shots. $310.

United (Williams) DELTA, 1968. With the death of founder Lyn Durant, United Manufacturing Company sold out to Williams, and the renamed Williams Electronic Manufacturing Corporation moved into the United plant. United's claim to fame in shuffles and bowlers was so strong Williams continued to make the machines under the United name. DELTA is a 4-player shuffle alley with features developed by United in their line ending SKIPPY made in 1963. $360.

Chicago Coin STEAM SHOVEL, 1956. By the '50s Chicago Coin was a copycat producer, with little original coming out of the firm. One of the exceptions to this rule was a kiddie game called STEAM SHOVEL. It was like playing in the sand on the beach, only it is in a covered box. It looks like a digger, but all you do is lift steel beams and pile them up. The more you pile the higher the score. Williams came out with a sand version called CRANE a few months later. $360.

Chicago Coin DRIVE MASTER, 1969. An upgrading of the DRIVEMOBILE machine made by International Mutoscope just before WWII (The Chicago Coin version was often called DRIVEMOBILE in the trade press), DRIVE MASTER offered "Windshield View" driving by rolling a drum with lighted road graphics behind a windshield in front of an automobile steering wheel. Drum is driven by a "gas pedal" on the floor. $265.

Midway GOLDEN ARM, 1969. Key design and engineering individuals of United Manufacturing formed Midway Manufacturing Company in Franklin Park, Illinois, in 1959. They grew rapidly and by 1967 added facilities in neighboring Schiller Park, another Chicago suburb. For GOLDEN ARM they went back to the old saloon Indian wrestling machines of the past and put the new version in a colorful modern cabinet. Return push is provided by a pneumatic dashpot. Strong stuff. $230.

United MIDGET ALLEY, 1958. Ball bowlers and shuffles were produced in a great variety of models. While a lot of fun, they also take up a lot of room. That fact alone has held back their collectibility and kept values exceedingly low. You get a lot of machine for the money. MIDGET ALLEY has the advantage of being short with all of the features of the more sophisticated ball bowlers, such as 2-player drum scoring, automatic pin setting and ball return. $310.

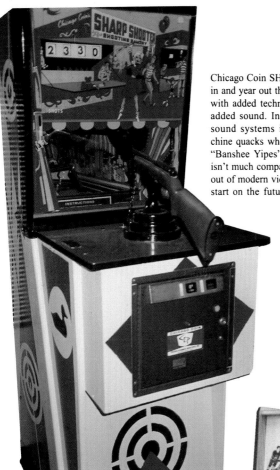

Chicago Coin SHARPSHOOTER, 1971. Year in and year out the gun games were improved with added technologies. SHARPSHOOTER added sound. In one of the first solid state sound systems in coin-op games, the machine quacks when a duck is hit, and sounds "Banshee Yipes" when a target is scored. It isn't much compared to the marvelous sounds out of modern video games, but it was a great start on the future. Drum scoring. $310.

Ramtek HOCKEY, 1973. When Atari, Inc. came out with the simulated ping-pong video PONG machine in 1972 it revolutionized arcade games. PONG was everywhere; airports, shopping malls, restaurants and clubs. It became the most widely copied game in arcade history with dozens of copycat producers. Ramtek Corp, Williams, and Chicago Coin rushed out with hockey variations in 1973, including PRO-HOCKEY, TV GOALEE and others in 2- and 4-player models. Big hits! $110.

Dynamo PRO-BUILT SOCCER, 1982. Other imports were German and Italian *Füssball* games, carving out a part of the dedicated mall arcade leisure market. Dynamo Corporation of Grand Prairie, Texas, started making the games in 1973, upgrading over the years with other American makers joining the fold. The game is basically the same as the Chester-Pollard PLAY FOOTBALL of the mid-1920s. What goes around comes around. Great game for home rec room play. $210.

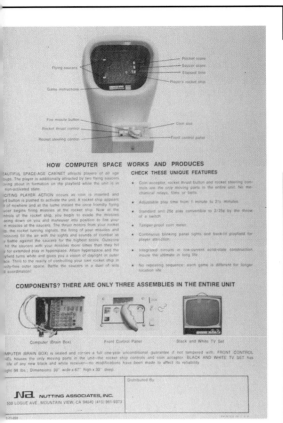

Nutting COMPUTER SPACE, 1971. The first video game. Nutting Associates, Inc. of Mountain View, California, made coin-op fortune tellers and COMPUTER QUIZ trivia machines. A young graduate electrical engineer named Nolan Bushnell brought a college computer lab game to Nutting, who made it in a fiberglass cabinet as COMPUTER SPACE. It sold okay, but Bushnell left to form his own Atari, Inc. and brought out PONG. Model is a young Yvette Mimieux before movie fame. $90.

Midway DELUXE SPACE INVADERS, 1979. Video games took over the arcade machine industry, led by Atari, Inc. in California and a suddenly emerging Japanese video game industry, with many produced under licence in the United States. Midway built their soaring success on Taito games. SPACE INVADERS of 1978 was the first mega hit, and established the genre. The DELUXE model of 1979 added color, more firepower, sounds and speed. Next came PAC MAN. $210.

Trade Stimulators and Counter Games

Not all coin operated gambling machines were payouts. The trade stimulators from the 1880s through the end of the teens, and the subsequent counter games of the 1920s through the 1950s, were just as chancey as their big payout brothers, but the prizes were minimal and the play was delicate and often intricate. The original trade stimulators started out as coin-op versions of the bartop games and dicers of the past, evolving into larger store pieces that gave a merchant supplied with them an advantage over one that didn't have the games on the counter. That lasted through the saloon years, and ended with the rise of Prohibition in 1920.

By 1920 the games had become a common adjunct to store and cigar counters, and so they survived their saloon origins. These were generally made in the post-WWI wonder metal, aluminum. The newer machines were called counter games, and were produced until the 1950s when the supermarket checkout counters replaced the Ma & Pa cash registers of the past, taking the location or counter space for the games with them. Trade stimulators and counter games are the only form of coin-op machines that have lost their place in time, and are no longer found in active usage. Scales are almost gone, but are coming back, so the counter games are alone in their demise.

Even though the counter games and their trade stimulator antecedents are chance machines, collectors need not worry about having the law on their side, because it already is. It is the payout slot machines, and the gambling machines, that have the legislators worried. That's not the case with the trade stimulator and it's later numerous offsprings, the 1920-1951 counter games. These soft-core chance machines do not make physical cash payouts, so they are eminently collectible in any state. But in actual usage they did qualify the player for a reward (or "award," as was common in the vernacular) if they "won" the game. Their heyday was in the 1897-1920 period when stores needed an advantage over the competition. So the merchants put in cigar and merchandise machines that played on a penny or a nickel (some, such as the cigar counter or cigar store WIZARD CLOCK machines, played on a dime) in which you got a number that said "1" (which meant you got your nickel's worth) or "2" (which means you got twice the value for your coin) or even more. The store that had one on the counter had an advantage over the store that didn't.

Later, in the '20s and '30s, counter games that featured the same fruit symbols as their larger slot machine cousins, or had other symbols or colored wheels and the like, carried on the tradition of getting more for your money, or at least as much (the latter solved by issuing a gum ball with each play in order to "give the player their money's worth"). All in all, while not hard gambling, they were on the fringe. In the case of the fruit symbol machines you have to assume that the slot machine collectible laws in some states may have application if the symbols are mentioned in their anti-crime laws. In states where such laws aren't on the books, the display and usage of such machines may need to be tested in court, but you are probably safe enough. Unless some officer of the law wants to run for election or something and hits on victimless crimes. But even there, once tested, it is difficult to believe that a modern judge or jury would hold you accountable for having a gambling machine on your premises. They couldn't make such laws stick with the counter games of the past, so they can hardly do so today.

The charms of the early trade stimulators are that they are old, going back to 1887, and that they offer a great variety of formats: dice, roulette, wheels, spinning pointers and all sorts of other clever mechanical and moving contrivances. These games are quite often very valuable because of this, and also because they tend to be rare. Trade stimulators were very cheaply made in most cases, and when their days of service were over they were usually tossed out. That makes the survivors significant. The most desirable are the cast iron cabinet games produced in the early 1900s, cheap then but regarded as a sought after industrial art today.

McLoughlin (Winchester) GUESSING BANK, 1878. First coin-op gambling machine made in the United States. Patented as a toy bank, likely to get around the laws preventing patents for gambling devices. The front says "Pays Five For One If You Call The Number," so that's clear enough. Payoffs were in drinks or cash by a bartender. Inventor Edward S. McLoughlin of New York City sold the rights to Smith, Winchester & Company in South Windham, Connecticut, who made it. $5,520.

When Prohibition became the law of the land, one of the phenomena of the 1920s was the advent of the counter game, or miniature slot machine. Unlike their earlier trade stimulator cousins they looked a lot like the automatic payout slot machines of the day. And unlike their larger brothers they didn't pay out but had their awards delivered over the counter by the shopkeeper based on the winning symbols on the reels. But you did get a gum ball for your money which theoretically took the chance element out as you got something for your payment. Among the early examples were the elaborately metal faced center pull Caille JUNIOR BELL, Caille FORTUNE VENDER, and Silver King BALL GUM VENDER counter games, all made between 1926 and 1930.

These pseudo slot machines were followed by even smaller models that looked for all the world like miniature payouts, only they didn't pay. One of the most popular was the PURITAN BABY BELL, a particularly hot non-payout counter game format from the late 1920s well into the 1930s. It was created by the J.M. Sanders Manufacturing Company on west Lake Street in Chicago in 1928, and was sold under private labels to about a dozen other so-called manufacturers well into the '30s. But Sanders produced the game, no matter whose name seems to be in the metal. Over a hundred variations exist.

One of the major marketers of the PURITAN BABY BELL game was the Midwest Novelty Manufacturing Company, selling it between 1928 and 1931. Midwest sold models in plain, gumbal (Note the contraction of the words "gum" and "ball" into one simple, workable word. The vending machine professionals liked to do things like that), fruit, number and cigarette reel strips, and even had a money paying jackpot model. After 1931, the firm dropped the machine, changed it's name first to Lion Mfg. Co. and soon thereafter to Bally, to go on to become a major producer of pinballs and slot machines.

Following this semi-slot trend in the early 1930s, there was a move toward clandestine counter games to get around the growing body of law against the machines, or at least to fool the fuzz when they came into a joint to check. The two largest counter game makers, Groetchen and Daval, and a smaller producer named Bennett, made wooden cabinet poker and cigarette reel machines that looked like the then newly fashionable small tabletop radios. The typical Daval SMOKE REELS was introduced in the summer of 1938. Playing for a penny, if you got 3 alike you won a pack of cigarettes; 5 packs for 4 alike, and 10 packs for all alike. The latter was all but impossible, for many of the machines didn't have all 5 symbols on the reels, or sometimes the merchants blocked them out.

One of the problems with trade stimulators and counter games is the wear they experienced when they were in operation. More often than not the original award card paper is in bad shape, or missing, a similar situation faced with the decals used on the cabinets and glass. Fortunately for collectors much of this paper and software has been reproduced to replace pieces that have been degraded or worn. You can get new reel strips, and new award cards, from Bill Whelan, P.O. Box 617, Daly City, CA 94017. He has a price list for all his repro material, so ask for it.

McLoughlin (Winchester) PRETTY WAITER GIRL GUESSING BANK, 1880. If there was any question about the saloon placement of the GUESSING BANK, get a gander at this lady. Hardly a child's toy bank. It's the most lurid trade stimulator ever made. Only one known; in a "secret" collection. It's probably the most sought after machine in the country. The casting says "Pays Five For One If You Call The Number." The name denotes a barmaid of questionable virtue. $15,050.

Bishop THE TRIOGRAPH, 1889. Very early cigar trade stimulator that doubles as a peep show. Drop in a nickel and the pre-wound clockwork mechanism pops the dice. You also get a chance to look at the racy pictures in the viewing glass. If the four dice total 4 you get 4 cigars, with lesser amounts for easier tosses, most times none. But you saw the picture. Another nickel, another picture. Made by Charles C. Bishop & Company in St. Louis, Missouri, a lively river city. $4,510.

Weston SLOT MACHINE, 1892. The popularity of bagatelles and other pre-pinball marble games was mechanized as a cigar trade stimulator at least once, with the ball rolling Weston SLOT MACHINE duplicating the action of the popular twin cup die tossing Clawson AUTOMATIC DICE of 1890. The award card reads "Add together the numbers of the pockets into which the balls fall," with cigar awards based on the totals. Only one of these machines is known. $4,510.

Excelsior EXCELSIOR, 1890. The horse race was one of the first amusements to be mechanized, based on imported "French Race Games." Leading maker was the Excelsior Race Track Company of Chicago. Others were Flour City Manufacturing Company in Minneapolis with an 1889 game; Henry Behn in Union Hill, New Jersey, in 1889; Chicago Nickel Works in 1889; Henry Cordray in Brenham, Texas, in 1886 and others. They were so much alike they all sued each other, which came to naught. $3,215.

Weston SLOT MACHINE, 1892. The clockwork mechanism is automatically activated once a coin is dropped. The lift rod tilts the two bagatelle playfields back and moves the four steel balls to the top. Then the pinfields are lowered and the balls bounce back into the scoring channels. Getting the highest four "1" score requires all four balls to stack two high in the two center channels. Weston Slot Machine Company was located in Jamesville, New York, just outside of Syracuse.

Automatic Machine AUTOMATIC DICE, 1892. The easiest game to mechanize was dice tossing, so early dicers were popular. Automatic Machine Company of New York City took the concept a step further by popping the dice under two glass domes; a small center dome for popping and a larger surrounding dome to prevent manipulation of the popping dome and dice. Automatic Machine was one of the earliest line producers, making card drop machines as well as dicers. $1,210.

Lighton DICE SHAKER, 1892. The same year that the John Lighton Machine Company of Syracuse, New York, made their landmark 3-For-1 automatic payout SLOT MACHINE they made this dicer paying off in free cigars. The flat topped glass was an innovation as it eliminated the high cost of special dome breakage. Its replacement is an ordinary beer glass stocked by most saloons, placed upside down. Brass corners protect the cabinet. $3,015.

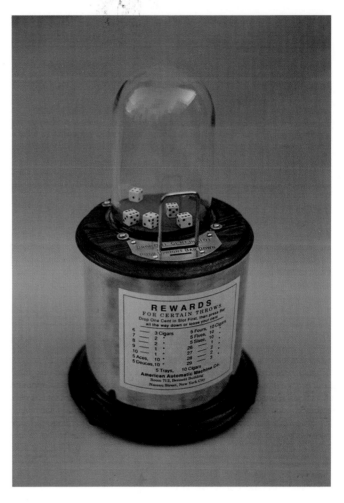

American Automatic AUTOMATIC DICE SHAKING MACHINE, 1893. Late 19th century dicers aren't common, but the one most frequently found is the AUTOMATIC DICE SHAKING MACHINE with its loop or plunger, suggesting that this was a machine that was in use everywhere. The dice get a great pop once the mechanism is tripped, with the lever pushing down on an eccentric cam that rides over and then snaps the bottom disc. American Automatic Machine Company was in New York City. $1,410.

Automatic Manufacturing AUTOMATIC DICE, 1893. With so many dice machines being made it took something exceptional to stand out and be effective. American Manufacturing Company of New York City produced the patented products of William R. Pope, with exclusive features such as the heavy duty knob flipping of AUTOMATIC DICE, getting away from the physical slamming of the delicate plungers of other machines. $1,510.

Western Weighing DIAL, 1893. Made by Western Weighing Machine Company and sold by Charles T. Maley Novelty Company, both of Cincinnati. DIAL is a no blank (meaning you always got something for your money) development of a coin drop called NICKEL TICKLER. The coin drops down, spins the dial in the center and lands in one of the pockets below. 2, 4 or 6 awards two cigars, all others one. If the pointer number on the dial matches the numbered chute you get five cigars. $3,020.

Western Automatic IMPROVED ROULETTE, 1894. Typical bartop and cigar counter trade stimulator in the late 1890s, made by a variety of producers between 1892 and into the 1900s. IMPROVED ROULETTE models with better wheel spin, improved mechanism and graphics were made by Clawson in Newark, New Jersey; Chas. T. Maley, Leo Canda Company and Western Automatic Machine Company in Cincinnati; Mansfield Brass Works in Mansfield, Ohio and others. $760.

Sittman & Pitt LITTLE MODEL CARD MACHINE, 1894. Increasingly card machines with spinning reels replaced the dicers and coin drops, until they became the major class of trade stimulators in saloon, restaurant, cigar counter and other primarily male locations. The firm of Sittman & Pitt of Brooklyn, New York, had patents on card drop and card roller versions. They also produced the machines private label, with Samuel Nafew Company of New York and Chicago the major buyer. $1,815.

Canda GIANT CARD, 1894. By the middle 1890s the largest producer of trade stimulators in the country was Leo Canda Company in Cincinnati, Ohio, making their own versions of the MODEL CARD MACHINE. Then they got an idea. Why not make them tall enough for a whole saloon to see the play. GIANT CARD was born, with GIANT POLICY, GIANT DICE and GIANT ARROW models with different reel symbols. To give you hope for discovering old machines, this only known example was found in New England early in 1994. $5,520.

United States Novelty JOKER, 1894. If you thought pre-1900 machines weren't colorful, take a look at this. The name of the wheel designer and design patentee is at the top of the wheel: George E. Stoneburner. This is one of those lucky machines that is blessed with a maker's name. Two pointers spin to fill out a hand of cards, with awards in cigars given by the bartender based on the result. Clockwork mechanism is wound up when needed. Only one known. $3,210.

Barnes CRESCENT (CIGAR WHEEL), 1897. The idea spread, but not far. Monroe Barnes Manufacturer was located in Bloomington, Illinois, the next large town north of Decatur. They all must have known each other, and they all sold major national markets. The Barnes CRESCENT CIGAR WHEEL is boxed and attractive, mostly because of the lady on front. CRESCENT has a double printed wheel, usable as a cigar trade stimulator and card wheel. Only three are known. $2,010.

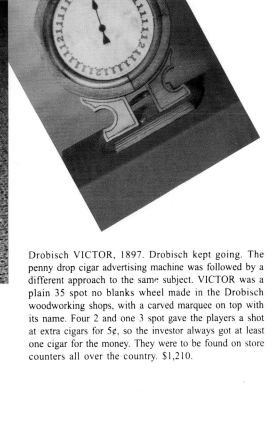

Siersdorfer COIN TARGET BANK, 1894. One of the very first of the boxed gun games, made by M. Siersdorfer & Company, Cincinnati, Ohio. The idea came from England and Germany where a similar game was produced from the early 1890s into the 1920s. Siersdorfer was a saloon equipment supplier who made the game to entertain the bar "Boys." Put your coin in the gun, shoot it, and if it goes into one of the holes you get the corresponding number of cigars indicated on the award card. $4,525.

Drobisch VICTOR, 1897. Drobisch kept going. The penny drop cigar advertising machine was followed by a different approach to the same subject. VICTOR was a plain 35 spot no blanks wheel made in the Drobisch woodworking shops, with a carved marquee on top with its name. Four 2 and one 3 spot gave the players a shot at extra cigars for 5¢, so the investor always got at least one cigar for the money. They were to be found on store counters all over the country. $1,210.

141

Drobisch ADVERTISING REGISTER, 1896. The prairie farming center of Decatur, Illinois, became a center of trade stimulator development. It all started in 1895 when the Decatur Fairest Wheel Company was formed to make the FAIREST WHEEL, the trade stimulator that took the genre out of saloons and put them in stores. Local competition was immediate. Newly formed Drobisch Brothers & Company came back with a penny drop that could put advertising on its face: ADVERTISING REGISTER. $1,210.

Merriam Collins PEERLESS ADVERTISER, 1897. Regarded by some (me included!) as the finest late 19th century advertising cigar wheel, the PEERLESS ADVERTISER, also called THE CODE based on the cigar it advertises, was made by Merriam Collins & Company of Decatur, Illinois. Only this time it took a foundry to make the machine, not a wood shop. Drop a nickel in the top chute and the roulette wheel below spins, with the steel ball coming to a stop in a 1, 2 or 3 hole. $3,215.

Drobisch THE LEADER, 1897. Basically the same machine with a different advertiser and, more significant, a different wheel. Play is 5¢, the cost of a cigar. The front says, "Drop a nickel in the slot. Try Your Luck." But this time you could miss, for in the 48 position wheel 32 are a zero, no cigar. There are some 1, 2 and 5 spots, and a single 10, but the risk was high and heavily favored the house. But 10 cigars for a nickel? It got played. $1,410.

Sun THE BICYCLE, 1898. First made by Waddel Wooden Ware Works Company (called the "Four Ws") in Greenfield, Ohio, a former partner started his own Sun Manufacturing Company in town, also making THE BICYCLE. Machine is found with both nameplates, etched in ivory on the front to the left of the handle. Drop the coin in at upper right, it rings the bell, and you crank the handle to spin the bicycle wheels. Total of numbers on the edge when they stop determines wins. $3,210.

Canda PERFECTION CARD, 1898. With over thirty trade stimulators behind them, Leo Canda Company finally created their classic. Called PERFECTION CARD, and nicknamed "Round Top" for its design, it became the most widely produced card machine of all time as the LITTLE PERFECTION, made by Mills Novelty (who bought out Canda), Caille Bros., Watling and a host of other producers. This machine was still selling in the early 1930s. $660.

Brunhoff DIAMOND EYE, 1899. There were a lot of ways to make a cigar wheel. Wild and outlandish ideas were tried. In all probability many of them have yet to be discovered. One of the oddest is the DIAMOND EYE by Brunhoff Manufacturing Company of Cincinnati, Ohio. The top carries advertising for Diamond Eye cigars. It spins under the glass dome when a coin is dropped and lever pushed down. Numbers at the top indicate the cigars won. "Pat Apd For" on side. $3,820.

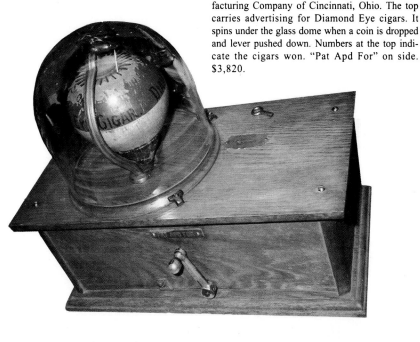

Decatur FAIREST WHEEL NO. 3, 1899. By 1897 the Decatur Fairest Wheel Company had become the Decatur Fairest Wheel Works of Decatur, Illinois, with continuing development of the namesake cigar and store discount wheel. One model gave 1, 2 or 3 cigars for a nickel; another gave discounts based on the percentage marked on the wheel. FAIREST WHEEL NO. 3 is smaller than the original model, and has a visible cash box so the owner can check the level of play. $710.

Mills Novelty LITTLE MONTE CARLO, 1898. Quite possibly the first acquired machine made by the newly formed Mills Novelty Company in Chicago, the LITTLE MONTE CARLO was originally made by the National Manufacturing Company in New York City in 1897, soon selling out to Mills. It was a major advancement over the wooden roulette games previously made. 5-way, you pick your color to win. Rare and valuable. $6,530.

Brunhoff SPINNING TOP, 1899. Somewhat more sophisticated and cleaned up, the final patented version of SPINNING TOP is smaller. It was used for trade stimulation and earned awards in trade tokens. The top is marked off in numbers, with certain numbers providing larger awards. Put coin in the coin head at top right, slide it over, push lever at bottom, top spins and you wait for it to stop at a good number. $3,410.

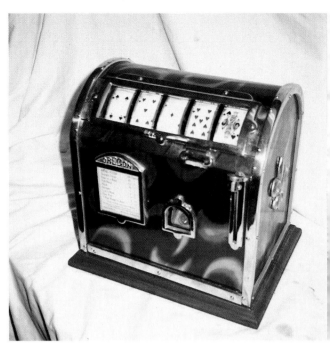

Wheeland OREGON, 1902. There were reels, wheels and drop cards to deliver hand at poker for a coin. More often than not the cards were smaller than actu playing card size, or even as small as a Little Duke set. But the Wheeland Novel Company of Seattle, Washington, took the next step. They used actual full size card on their reels so that the visibility is great, providing big cards on a small machin Only one of these is known. $7,510.

Mills Novelty MANILA, 1899. From the sublime to the super. Compare to the Siersdorfer COIN TARGET BANK of 1894. The Mills Novelty MANILA is four times the machine; in size, and performance. Created by the Chicago firm of Paupa & Hochriem in the summer of 1899, by winter Mills was making it. Shoot nickels, and if they go into the target holes you get paid anywhere from 1 to 10 trade tokens from four payout mechanisms. Two known. This one was found in Tasmania in 1993. $10,050.

Kelly FLIP FLAP, 1901. The FLIP FLAP (also called the LOOP THE LOOP, b the original name is the accurate one) was made by the Kelley Manufacturing Co pany, Chicago, Illinois, a cigar distributor. The coin goes down the brass chu whips up some speed as it makes a loop and comes zipping out at the left side bounce back to a 1, 2 or 3 scoring hole to win some cigars. Hard to control as t speed was virtually always the same. $2,210.

Bryant ZODIAC, 1902. Often it has been difficult to identify machines long after they were produced. When the ZODIAC first showed up it looked like a Caille Bros. machine, perhaps an early WASP. Then a patent was discovered that suggested it was made by Wain & Bryant Company of Detroit, a company Caille Bros. took over. Then a machine showed up with the maker's name Bryant Pattern & Novelty Company of Detroit in its casting, presumably also bought out by Caille Bros. $7,510.

Mills Novelty RELIABLE, 1903. The most elaborate and possibly the best looking drop card poker player. The Mills RELIABLE gets two coins from the player if they feel they have a good start on a winning hand. Play the right side first and get a poker hand. If its good, or just close, a second coin on the left hand side spins the single reel, with the hand pointing to the card that was drawn. If you get a better winning hand you get paid off in cigars. Only one known. $7,510.

Mills Novelty KING DO-DO, 1903. Mills Novelty was known for its elaborate case castings for its cast iron card machines. KING DODO is one of the best. Well endowed ladies grace the sides, with an artistic script on the front. The machine was named after a popular stage show of the period. It came in both nickel and penny 5-way and nickel, dime and quarter 3-way models, with the 3-way shown. The windows below the coin slots show the last coin played, so no slugs. $3,810.

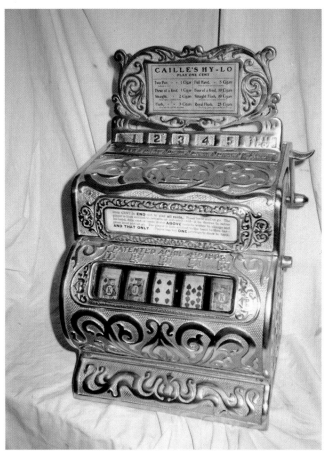

Caille Bros. HY-LO, 1904. In a few short years the cast iron card machines evolved from plain wooden cabinets with cast trim to painted iron and anodized copper countertops and then to dazzling bright nickel plated heavyweights. HY-LO is The Caille Brothers Company version of a draw poker machine that lets the player fill in a hand. A coin at the far right, and handle pull, spins all reels. To fill out add coins to the numbered slots and pull again. $4,410.

Mills Novelty HY-LO, 1904. One of the continual charms of vintage coin machine collecting is to see how the different companies handled similar sales and machine operation situations. The Mills Novelty HY-LO is essentially the same machine as the Caille Bros. HY-LO, except it has turned everything upside down. On the Mills version the coin slots are below the reels, just the opposite of the Caille. The Mills handle is low, the Caille high. Same, but different. $4,410.

Caille Bros. WASP, 1904. You can see that Caille Brothers got the idea for the WASP from the Bryant Pattern & Novelty Company ZODIAC, plus some evolution. WASP is a 5-way color wheel that works the same way as floor model slot machine of the period, except it doesn't pay out. Two coin heads were offered. On one you got a cigar for every play, maybe more, provided you asked for it. On this model you played a color and got cigars or cash accordingly, most often none. $4,210.

Mills Novelty DRAW POKER, 1904. All of the major trade stimulator makers had a DRAW POKER machine, differing only in details. The machines were advanced developments of the older and much smaller Sittman & Pitt card drops, adding elaborate plunger stops for each reel. You generally got two spins for a nickel. Play, push down the hold buttons for the cards you want to save, and play again. Payoffs on the final result. Mills name in base. $4,010.

Mills Novelty THE TRADER, 1904. The buyouts in the coin machine business were endemic in the first decade of the 20th century. The larger companies were on the lookout for new ideas, and bought out smaller companies as fast as they came up with machines. Mills Novelty bought out the Royal Novelty Company in San Francisco, makers of the ROYAL TRADER. Mills gave the machine one of its fabulous copper cabinets and continued making it as THE TRADER. $3,210.

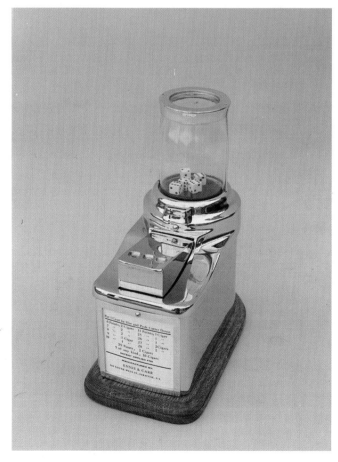

Ennis & Carr PERFECTION, 1904. PERFECTION was probably the most used name in trade stimulators, applied to card machines, wheels, coin drops and dicers like this one. The maker was Ennis & Carr located in Syracuse, New York. When these machines were first found they were thought to be a lot newer. But research proved otherwise, with these 1920-ish and even '30-ish looking dicers made in the early 1900s. One clue is the holes in front; they're cigar cutters. $2,610.

147

Caille Bros. REGISTER, 1906. Slot machines were often compared to cash registers, similar heavy cast iron devices that also handled money. It was in trade stimulators that the comparison was strongest. Caille Brothers Company played up to that thought with their REGISTER cigar machine, even to the point of having a wooden cash base. It plays like a three dice game, with the sum of numbers on three reels adding up to winning or losing counts. Rare. $4,210.

Cowper DRAW POKER, 1906. Until the summer of 1993 no one had any idea that the Cowper Manufacturing Company of Chicago had the DRAW POKER machine in their line. It was never advertised or cataloged. At least not in the vintage paper found to date. Then one of these machines showed up in the middle of Wisconsin, and history was revised. It made a fast trip through three or four owners before it ended up in a permanent collection. $5,020.

Griswold STAR, 1905. The M. O. Griswold & Company of Rock Island, Illinois, was a very early trade stimulator maker, cataloging machines as early as 1892. Their most famous trade stimulator was an 1893 counter wheel called the WHEEL OF FORTUNE, noted for its half visible wheel. In 1905 they became the Griswold Manufacturing Company, and continued making a version of the earlier machine with a fully exposed wheel called STAR. Popular well into the '30s. $660.

Hamilton DAISY, 1907. The trade machines that appeared in stores and candy, cigar and news counters provide a rich field of hunting for latter day collectors. Hamilton Manufacturing Company in Hamilton, Ohio, made one of the most popular. DAISY is a no blank, with players getting 5¢ or 10¢ in trade for each nickel played, depending where the coin landed. This model was nicknamed "Bread Loaf." A later slab sided version was called "Diamond Top." $310.

Opposite page photo at right: Caille Bros. BON TON, 1907. Trade machine for stores, with payoffs in merchandise based on trade tokens ranging from 2-1/2¢ to 30¢. Half cent sales were common in the early 1900s (saloons sold mixed drinks for 12-1/2¢!). On BON TON if you got "1" on the first reel, you got paid with a trade token for 2-1/2¢. If you got "2" on the second reel, "3" on the third and so forth the token payouts were higher. Tokens are dispensed in the cup on the side. $4,815.

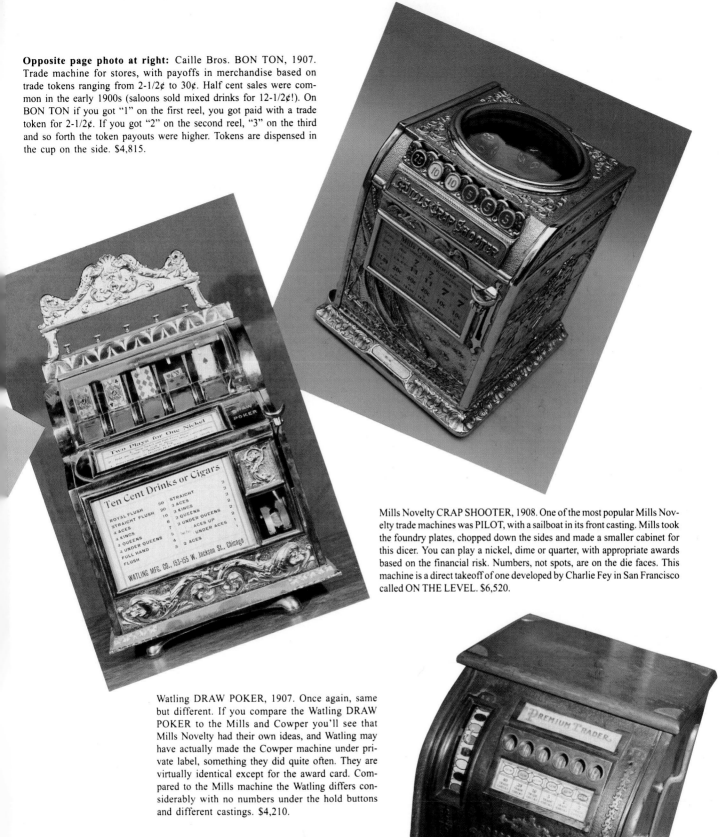

Mills Novelty CRAP SHOOTER, 1908. One of the most popular Mills Novelty trade machines was PILOT, with a sailboat in its front casting. Mills took the foundry plates, chopped down the sides and made a smaller cabinet for this dicer. You can play a nickel, dime or quarter, with appropriate awards based on the financial risk. Numbers, not spots, are on the die faces. This machine is a direct takeoff of one developed by Charlie Fey in San Francisco called ON THE LEVEL. $6,520.

Watling DRAW POKER, 1907. Once again, same but different. If you compare the Watling DRAW POKER to the Mills and Cowper you'll see that Mills Novelty had their own ideas, and Watling may have actually made the Cowper machine under private label, something they did quite often. They are virtually identical except for the award card. Compared to the Mills machine the Watling differs considerably with no numbers under the hold buttons and different castings. $4,210.

F. W. Mills PREMIUM TRADER, 1917. The name Mills should ring a bell. Same family, different man. Frank W. Mills was the brother of H. S. Mills of the Mills Novelty Company. Off and on Frank had his own companies. F. W. Mills Manufacturing Company was located in Chicago. PREMIUM TRADER is a revamp of the Mills Novelty CHECK BOY payout slot, only it is a plainer store machine, in a wooden cabinet, and pays off in trade checks, the "premiums" of the name. $3,415.

Page SALES INCREASER, 1909. A stimulator is intended to stimulate trade, usually with a coin machine. So the SALES INCREASER product of the Page Manufacturing Company of Chicago fits the bill. It actually fits over a cash register, certainly a form of coin machine. Ring up a sale, push in the drawer and the spinner twirls. If it matches the sale you get it free. These are rare. Do cash register collectors have them? Otherwise, where are they? $1,610.

Fey TRIPLE ROLL, 1925. Chas. Fey & Company of San Francisco never threw anything away. Certainly not foundry patterns. He took his circa 1907 dice game ON THE LEVEL and put a roulette wheel on its top, but kept the original name on the cabinet casting. The new game was called TRIPLE ROLL in Fey advertising flyers, only the award card said SKILL ROLL. If that isn't confusing enough, he kept the old dice symbol "Read 'Em" arch across the top to prevent slapping.

Banner LEADER, 1922. Prohibition changed the mores of a nation. Booze became the pot of its era; illegal but indulged. Trade stimulators became counter games for stores and checkout counters. Banner Specialty Company in Philadelphia, Pennsylvania, had long routes of commercial locations. So they made their own machine, a variation of the Mills LITTLE DREAM penny drop. Land in "G" you get a stick of gum. Land in 1, 2 or 5 and you get that amount in trade. Fairly common. $360.

Matheson KITZMILLER'S AUTOMATIC SALESMAN, 1925. A complicated and unique coin target machine was created by the Matheson Novelty & Manufacturing Company in Los Angeles, California, in 1924. They called it their ORIGINAL AUTOMATIC VENDER. A year later this version showed up. What it automatically sold was gumballs, vended at the side by pushing the plunger, while it gave you a shot at more for your money by making certain targets. So, who's Kitzmiller? $460.

Mills Novelty PURITAN, 1926. With Prohibition the law of the land, many old saloon machines were converted to commercial usage. The cast iron Mills Novelty PURITAN of the early 1900s got a new aluminum cabinet with large reel windows showing the numbers in red, white and blue. Another model used the fruit symbols of an automatic payout slot machine and picked up the name PURITAN BELL. Only these machines didn't pay off; you got your rewards in trade. $610.

Williams THE ADDER, 1925. When Charlie Fey allowed friends in Fond du Lac, Wisconsin, and others, to produce his counter games, his patterns made the rounds. One of the molders, Williams Manufacturing Company in Minneapolis, Minnesota, made a modified version of TRIPLE ROLL and called it THE ADDER, allowing a fill-out spin for a nickel, dime or quarter. The machine and company names were in the castings, although the confusing SKILL ROLL paper was retained. One known. $4,210.

Caille Bros. BALL GUM VENDER, 1926. Caille Brothers Company could read the future as well as the rest of them, and created their own non-payout "Baby" Bell machine with fruit reels, vending a ball of gum with each play. The BALL GUM VENDER became the JUNIOR BELL "Style 2" when it got an all aluminum front in 1928. It was also made by (or for!) Silver King in Indianapolis, Indiana, and Superior Confection in Columbus, Ohio, with unique fronts of their own. $900.

Ad-Lee TRY-IT BALL GUM VENDER, 1927. The Ad-Lee Company, located in downtown Chicago, was a vending machine marketer known for their E-Z ball gum vender. Chance features made it close to a trade stimulator, and they finally went that way in 1927 with TRY-IT, a small square dice tosser in the crackle finish so adored in the late '20s. A larger cabinet version, shown, added gumball vending capability. Great '20s graphics on the front. $365.

Engel IMP, 1927. The idea of a miniature non-payout Bell machine led to games that went for straight cash play and pay, the latter over the counter. A true counter game. The Engel Manufacturing Company of New Holland, Michigan, produced one of the first of the western Michigan small Bells, with the IMP followed up in 1931 by the Benton Harbor Novelty Company in Benton Harbor, Michigan, to add payout capability. $710.

Exhibit Supply ARCADE DICE FORTUNE TELLER, 1928. Ostensibly a fortune teller, and theoretically an arcade machine, the Exhibit Supply Company of Chicago duplicated the Clawson AUTOMATIC DICE cigar machine of 1890, right down to the clockwork two cup dice toss. The principle difference is that the Exhibit Supply machine is largely made out of aluminum whereas the 19th century Clawson machine was produced in cast iron. Other than that, same thing. $3,210.

Wall HUSTLER, 1927. It's a wonder that southern California didn't become a major coin machine center in the late '20s and '30s as they had so many start up companies. But they closed just as fast. One quickly blinking star was the Wall Novelty Company of Huntington Park, California. Basically a five reel card machine, HUSTLER hid that concept by using number reels. But it didn't fool the authorities as evidenced by the playing card tax stamp at the left. $2,610.

Superior Confection BASE BALL AMUSEMENT, 1929. A classic example of taking the existing and making the new. Under the skin this is the Caille Bros. inspired FORTUNE BALL GUM VENDER of the Superior Confection Company of Columbus, Ohio. Superficially, it is a baseball game with a new cast aluminum front, scoring diamond marquee at the top, and baseball reels. When radio made baseball hot at the end of the '20s, base ball games blossomed. Two known. $2,010.

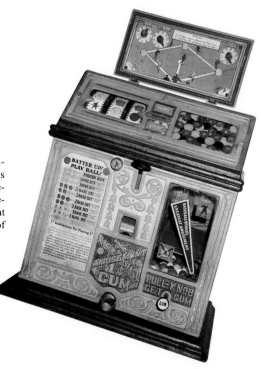

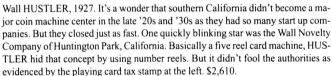

ey THE ACE, 1927. Edmund C. Fey, son of Charlie Fey of San Francisco, as also a coin machine maker. His THE ACE is a variation of the old -column dicer, only the columns aren't delicate tubular glass. They are uminum troughs covered by a flat glass. Clever man, Ed. Two throws for a ckel. Throw once, then push buttons to hold what you want to save, and ish the plunger for a second throw. The locations were former saloons, lling root beer and cigars during Prohibition. $3,420.

Superior Confection AUTOMATIC DICE TABLE, 1930. Prohibition or no, the '30s were years of hope. Everything was tried! Imagine adding coin control to a serving table? Superior Confection did it, and made three or four models of the device. Or is this furniture? In the AUTOMATIC DICE TABLE a coin slide actuates a dice popper in the center of the table. Other models included card shufflers and coin operated mechanical bridge scoring. Few found, all in Michigan. $410.

Field INDOOR SPORTS, 1932. Field Manufacturing Corporation of Aurora, Illinois, made a series of truly unique counter games. INDOOR SPORTS is a betting game for either poker hands or numbers. A header was added that said POKER-E-NO. Later renamed DING THE DINGER with a new marquee. Drop coin, turn knob at top, and the five weights are released and drop. Then you hit the finger flippers one at a time to lift the weights for cards or scores. Few known. $460.

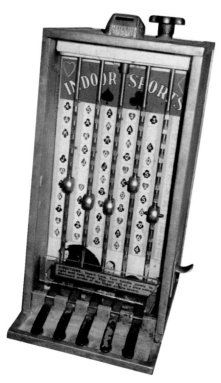

Pierce BUGG HOUSE, 1932. The deep years of the Great Depression were 1931-33, with no evident hope of a turnaround. A wave of entertainment devices made by new companies resulted, including pinball and hundreds of counter games. Pierce Tool & Manufacturing Company of Chicago made both forms of games, often unique. BUGG HOUSE has five rows of shooters, taking a coin each. Scores of a thousand and up earned counter payoffs. It's hard to hit across to the 500 squares. $460.

Groetchen POK-O-REEL GUM VENDOR, 1932. A coin machine industry subcontractor, Groetchen Tool & Manufacturing Company made their first house label counter game in 1930. By the end of 1932 they were one of the two leading counter game producers. The game that put them there was POK-O-REEL, a modern interpretation of the old card machine in a colorfully painted aluminum cabinet that added a gumball vender on some models. A new generation of players embraced the game. $410.

Silver King LITTLE PRINCE, 1933. Pierce Tool made a 3-reel fruit symbol game called WHIRLWIND that skirted the edge between the automatic payout slot and the counter game. Too small and light to stand up to slot machine usage, it remained a counter game. It's design was said to be based on the architecture at the 1933 Chicago World's Fair. A private label version with cigarette reels and a new front name casting was made for Silver King in Indianapolis, Indiana. Rare. $2,410.

Great States FOOT BALL PRACTICE, 1932. A new counter game and pinball maker, Great States Manufacturing Company was located in Kansas City, Missouri. Taking a tried and true format called TARGET PRACTICE, made by over a dozen producers over the years, Great States gave it a football field. New game! Scoring holes were added to the center of the playfield. It became known as the "Football Pin Target." Not particularly successful. $560.

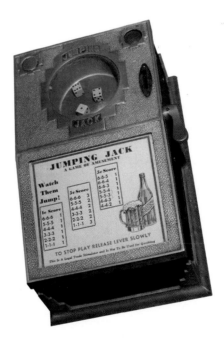

New Era JUMPING JACK, 1933. A beauty it ain't, but JUMPING JACK has a lot going for it otherwise. It's a beer machine! With the November 1932 election of Franklin Delano Roosevelt and his promise to end Prohibition, Congress gave him what he wanted and beer was back on sale in May 1933. JUMPING JACK with a foaming stein on its award card came out in July. New Era (even the company was named after the repeal of Prohibition) Manufacturing Company was located in Chicago. $365.

Yendes REEL-O-BALL, 1932. Yendes Manufacturing & Sales Company¾-a division of Yendes Service, Inc.¾-a Dayton, Ohio jukebox distributor, promoted their baseball themed REEL-O-BALL as a new idea. The 5-way single reel machine has a gumball vender and shows baseball scores on the reel at far left. The mechanisms are actually from Paupa & Hochriem, Mills Novelty and Caille Bros. models of the single reel ELK, SPECIAL, TIGER and similar machines from the early 1900s. $1,210.

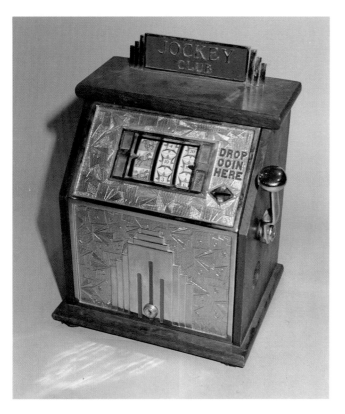

A. B. C. Coin JOCKEY CLUB, 1933. Another added machine starter, A. B. C. Coin Machine Company, Inc. was formed in Chicago in 1933 to take advantage of the many new tavern locations that would need counter games. JOCKEY CLUB has a twist. Looking much like a small slot machine with three reels, the symbols were of horses with a numbered horse on the first reel, numbers on the second reel that had to match the first to win, and payout odds on the third reel if it did. $620.

Daval ULTRA MODERN DAVAL GUM VENDER JACKPOT, 1933. Daval Manufacturing Company of Chicago started out as A. S. Douglis & Company. Then, in 1932, with a new machine to make, the company was renamed after its two partners, Dave and Al. The DAVAL GUM VENDER came in a maze of models, short (on the ULTRA MODERN Deco base), REGULAR or with jackpots (that had to be unlocked to release when 3 bars were hit) and others. By the end of 1933 they were neck and neck with Groetchen. $920.

Great States HAPPY DAYS, 1933. Not one to let a good idea get away, Great States took their FOOT BALL PRACTICE format and made a coin shooter target game that looked like a stein of beer. The theme song for the Democratic National Convention in 1932 was "Happy Days Are Here Again," meaning Roosevelt. With Repeal, the song meant the return of beer. The new tavern locations loved HAPPY DAYS as it advertised the fact they sold beer. Prime piece, and rare. $2,015.

Great States BUCK-A-DAY, 1933. Great States was a busy outfit in 1933. In fact, the whole country was as thousands of new taverns opened for business, requiring all the outfitting and equipment needed by a bar. That alone made the national business picture look better. Smart politician, that Roosevelt. BUCK-A-DAY is a six slot dicer that allowed multiple players. Imagine the arguments over who and how the dice were flipped. $310.

Jennings CARD MACHINE, 1934. Even the big slot machine makers saw the business opportunities in counter games, so they added them to their lines. The early O. D. Jennings & Company offerings made in Chicago were plain and unimaginative. CARD MACHINE looked like something out of the saloon era in its plain wooden cabinet. And it played the same way; old fashioned. A variety of models were made in this dull cabinet: CARD MACHINE, later called POKER GAME, and 21 BLACKJACK. $410.

Jennings REBATER, 1934. The backup model of the Jennings LITTLE MERCHANT, this time in fruit reels. In the REBATER you do the same thing and set up the win, pull and try and match the two sets of symbols. But it isn't cigarettes or gum you're shooting for. It's a rebate, or money back. Over the counter. But to be sure you got something for your money, if you got the symbols on the award card you get to read your fortune. Again, too complicated. $715.

Groetchen 21 VENDER, 1934. New counter game development was nothing short of amazing once thousands of new tavern and roadhouse locations were available. Groetchen (and Daval) outdesigned and outengineered the competition until most of the newer, smaller shops closed. 21 VENDER is very sophisticated, simulating blackjack. Five reels spin, but the last three are shuttered. You see two cards and can hit up to three times to win on 19, 20 or 21. Over 21 and you're out. $460.

Jennings LITTLE MERCHANT, 1934. Jennings wised up fast, and came up with some new ideas. LITTLE MERCHANT is exceedingly clever. It came in either cigarette or number reels because it was a selling machine, smokes or trade. The player sets up the win by dialing the three reels at the bottom with the knurled wheel sets. Then plays. The handle pulls locks the selection, and if the reels match top and bottom you walk away a winner. But it was too complicated, so it's rare. $560.

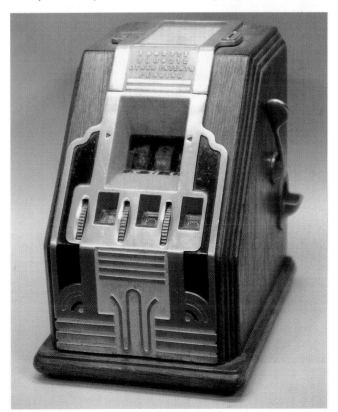

Kentucky Gum LUCKY BOY, 1934. Collectors get excited over "unadvertised" games. These are machines that were not promoted in the trade press, with no ads that reveal when they were made or what they looked like. It takes a surviving machine to reveal its existence. LUCKY BOY showed up in 1994. By a fluke of luck it had been briefly mentioned in an obscure Texas trade publication in November 1934, thus a date. Maker is Kentucky Gum Company of Louisville, Kentucky. $620.

Fort Wayne LITTLE JOE, 1934. The LITTLE JOE dicer made by the Fort Wayne Novelty Manufacturing Company of Fort Wayne, Indiana, looks like something out of the 1890s. Except for that outrageous Deco base. In its short two summer lifetime the Chicago World's Fair did more to architecture and product design that any prior event except perhaps the fair of 1893, also located in Chicago. Over the counter payouts are indicated in dollars and cents, a perfect tavern piece. $560.

Rock-Ola HOLD AND DRAW, 1934. Even Rock-Ola Manufacturing Corporation in Chicago felt the need to get into post-Repeal counter games, making a number of them. Their most heavily promoted game was HOLD AND DRAW, a modern version of the old DRAW POKER machines. Five card reels and hold buttons, and two plays for each coin. Yet for all the promotion, and there was a lot, these games seem to be rare so they couldn't have gone over very well. $710.

Ohio (Groetchen) THE BARTENDER, 1935. The Ohio Specialty Company, a tavern equipment supply house located in Cincinnati, Ohio (where a number of the old saloon supply houses had been located), handled the Groetchen TAVERN STYLE A beer and STYLE B cigarette symbol counter games. Then they took the idea a step farther and had Groetchen make a run of the two models with a new header identified as THE BARTENDER, with Ohio Specialty having exclusive distribution rights. $515.

Superior Confection CIGARETTE, 1935. After years of aping other products, and making lashups with old components, Superior Confection finally got into the swing of making indigenous machines in their own factory. The result was a series of unique designs produced by no one else. The CIGARETTE has a sleek, modern look for 1935. It also has 20-stop reels just like a payout slot, which makes the machine run like a well tuned clock. $460.

Pace SUDS, 1935. Beer symbol counter games were the rage before pouring spot marketing of tap beer got so super sophisticated that the counter games no longer were needed to push the product. Pace was making a cigarette counter game with a gumball vending feature called HOL-E-SMOKES. The same basic cabinet with a new front name casting and beer reels became SUDS. An old "growler" take home beer bucket, gone with Prohibition, was among the winning symbols. $510.

Superior Confection CIGARETTE GUM VENDER, 1935. The gumball vending version of the Superior Confection CIGARETTE has a confection compartment so large it could probably run for a week or more even if the players kept pulling out the gumballs, which they usually didn't do. The machine is penny play only; it says so on the front castings. To get the gumball you could pull down the small "drawbridge" vending lip at the bottom center of the machine with each play. $520.

Pacific ELECTRIC SPINNER, 1935. The flat box dice popper was a standard format in the middle '30s for tavern and cigar counter dice games. Early on beer mugs were used in the graphics, but as drinking became commonplace they disappeared. ELECTRIC SPINNER, made by the Pacific Amusement Manufacturing Company, a Los Angles firm that moved to Chicago, pops the dice electrically. It is the same game as Pacific's P. D. Q. with a new top glass. $260.

Exhibit Supply RED DOG, 1937. There was more than one way to make a 5-reel card machine. Charlie Fey worked this one out and Exhibit Supply in Chicago made an advanced development. RED DOG is the name of an old poker game. Five discs spin behind small windows, stopping at card symbols. Push in the slide and the house card at left is dealt. Pull out the slide and your four cards are dealt. If you have two or more higher cards in the same suit as the house card, you win. $615.

Groetchen HIGH TENSION, 1936. Coin machine makers spin their patents and proprietary ideas off in as many directions as they can as long as a feature still has life. HIGH TENSION is a horse race themed game using the two plus three reels of the 21 VENDER of 1934, only this time they are all visible. The first reel sets up the odds; second shows position win, place or show; and three to five are alphabetical. If they spell a horse name shown on the card, you win. $560.

160

Jennings PENNY CLUB, 1938. After its early ugly counter games, Jennings went for style. And a lot of machinery. PENNY CLUB is a trade machine that gives a gumball with every play, and awards 15¢ to 60¢ in trade depending on what three cigarette packs you line up on the reels. The mechanism is as involved as a slot machine. The front award card graphics showing a green, yellow and red cigarette package is among the best looking in counter games, and much sought after. $560.

Bally BABY RESERVE, 1938. Once Bally was a counter game maker they made a series of models. BABY RESERVE has three number reels (fruit and cigarette reels also came with the machine) with awards of 2, 5, 10 or 20 depending on the numbers spun. Numbers ending with 22, 33, 44 etc. win 2; ending with 00 win 5; triples such as 222, 333, 444 etc. win 10; and 000 wins 20. Match the three digits in the window above and you win 50. But Oh Lordy, this thing is ugly! $410.

Keeney JITTER-BONES, 1939. One thing can be said for J. H. Keeney & Company of Chicago; they were usually very original. This game is a plug-in, so it's electrical. But more than that, it doesn't have a coin slide. You either pay the barkeep, or the location lets you play it for free (with losers buying drinks for the house, so they make out). It's the way that it operates that makes it so original. No coin. No switch. Just lay your palm on the glass and the dice pop. $250.

Western REEL POKER, 1939. Everybody had to get into the act. Some even twice. After Western Equipment & Supply Company of Chicago started making counter games, and then went into one of its many bankruptcies, coming out of it as Western Products, Inc., it got back into the same business. The Western Products games have a massive Deco look, but they are small. REEL POKER is a classic 5-reel poker machine, with the added plus that deuces are wild. $260.

Groetchen IMP, 1940. How could you possibly make a coin-op game smaller? And this thing vends gumballs yet. Once the Mills Novelty VEST POCKET payout Bell machine was made, an even smaller interpretation was envisioned as a counter game. Groetchen won the race with IMP, available in fruit (with a Black Cat!), number and cigarette reels, eventually settling into cigarettes. The game came back again after WWII as IMP, and in 1949 became ATOM. $160.

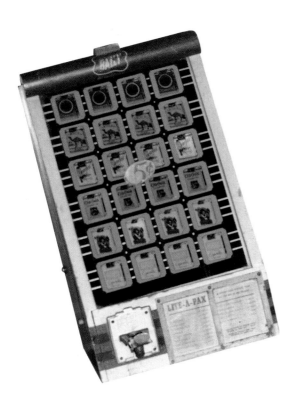

Bally LITE-A-PAX, 1937. Welcome to the future. The exciting all-electrical play of the automatic payout spinner consoles, and the lighted backglass scoring of pinball, re-made the world of coin-op entertainments. Bally Manufacturing Company of Chicago was deeply involved with both of these advancements, so they added them to the counter game. If the randomly flashing LITE-A-PAX lights up the four symbols of any cigarette brand, you win accordingly. 5¢ play in 1937. $310.

Daval CUB, 1940. The Daval answer to the Groetchen IMP was two machines. Actually one with a mask over the top. CUB is everything the IMP is, and about the same size. It also vends a gumball. Reels are fruit, numbers or cigarettes. But a matching model without the reel mask at the top, and a new front casting, came out as ACE, a 5-reel card machine. Including gumballs. Five reels in a palmfull of machine. Incredible. All three¾-IMP, CUB and ACE¾-sold like hotcakes. $140.

Daval AMERICAN EAGLE, 1940. Daval finally settled on their "Plain Vanilla" model. AMERICAN EAGLE established the basic Daval look for the pre-war (and immediate post-war) '40s. Symbols are fruit. The same basic machine in cigarette reels was called MARVEL, an off brand pack name at the end of the 1930s. STAR AMERICAN EAGLE has geometric symbols, including a star as the top winner. Gold Award and non-coin models of both AMERICAN EAGLE and MARVEL were made. $210.

Bennett TEST YOUR SKILL, 1941. Another unadvertised machine, and one that often stumps its owners because there are no names or dates on it. Paul Bennett & Company was formed in Chicago in 1938 to make counter games and jukebox needles. The counter games all look somewhat alike, in wooden cabinets with rounded edges. LUCKY PACK came out in 1938; DEUCES WILD in 1939; TOKETTE and DOUGH BOY in 1940; and TEST YOUR SKILL in 1941. It plays the same as the Bally BABY RESERVE. $260.

Daval BEST HAND, 1947. Counter games went right back in production after WWII. First as repeats of pre-war models. But then Daval came out with new machines. BEST HAND is a 2-player competitive game, with matching miniature "Whirlwind" playfields in which players try and beat the poker hand of their opponent. A baseball version was made called MEXICAN BASEBALL which scores in hits and runs. Then there was the "Girlie" game called OOMPH, which rates bathing beauties. $360.

Groetchen YANKEE, 1941. Groetchen's "Plain Vanilla" model for 1940 was LIBERTY, available in the same models as Daval's AMERICAN EAGLE and MARVEL. In 1941 Groetchen made a major line switch, using the IMP mechanism, and produced machines that had the front grilles of 1941 model cars. WINGS had cigarette reels; KLIX had numbers; POK-O-REEL had cards, and YANKEE came in both cigarette or fruit reels. POK-O-REEL, KLIX and YANKEE came back after the war in 1949. $210.

Arkansas Novelty 3-A-LIKE, 1947. Flat box dicer games of the 1930s began to look all alike. Exhibit Supply Company in Chicago made most of them. After WWII a rash of similar dicers began to show up with local operator names on them. Their rarity suggests only a few were made. Most are likely revamps of Exhibit Supply or other dicers, with a new top glass, because that's all it would take to make a new game. This is 3-A-LIKE of the Arkansas Novelty Company of Magnolia, Arkansas. $260.

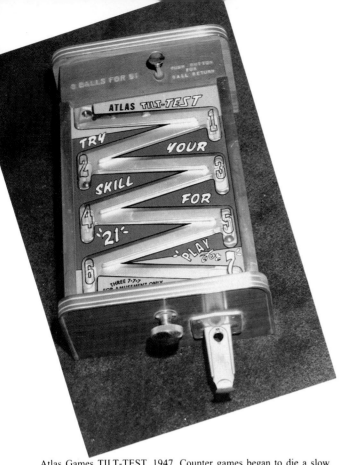

Atlas Games TILT-TEST, 1947. Counter games began to die a slow death after the war due to location loss. Cashier counters disappeared in stores as supermarkets took over, and taverns had much more sophisticated fare. Only a few minor games appeared. Atlas Games in Cleveland, Ohio, made TILT-TEST for bars, with 7-7-7 the winner. Very hard to score. Auto-Bell Novelty Company in Chicago took over the game in 1952. $180.

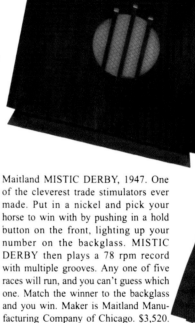

Maitland MISTIC DERBY, 1947. One of the cleverest trade stimulators ever made. Put in a nickel and pick your horse to win with by pushing in a hold button on the front, lighting up your number on the backglass. MISTIC DERBY then plays a 78 rpm record with multiple grooves. Any one of five races will run, and you can't guess which one. Match the winner to the backglass and you win. Maker is Maitland Manufacturing Company of Chicago. $3,520.

Unknown BOUNCERINO, 1952. Once the interstate highway program got underway in the '50s, and the shopping mall was where most Americans got their food, goods and services, counter games were dead. But out-of-the-way makers and locations still kept making and using a small number of the machines. Many are unidentifiable as no maker names appear. BOUNCERINO is the old pinfield format, offering up to $5 on a nickel play (with a blocking pin solidly over the $5 hole). $230.

Shelby LUCKY 7, 1978. If you didn't know this machine was made in the '70s you'd swear it was a '30s game. Some have been sold in the antique market with that understanding. They were made by an old operator, coin machine enthusiast and collector who thought that a legal vintage game idea might go over very well in local bars, and were marketed by the Shelby Agency, Inc., a real estate agent located in Lexington, Kentucky. He was right. Quite a few were made. $230.

Vending Machines

The odds are that if you have never before seen a vintage coin machine, or have just become aware of them, the first one you see will be a gumbal machine. In the 1970s they were faddish, becoming a staple in any college dorm room and even earned the honor of a Taiwan-made imitation that still sells to this day. They are colorful, cute, and highly collectible. Fact is, there are about as many gumbal (and other countertop) vending machine collectors as there are of pinballs, counter games and trade stimulators, although not as many as jukebox or slot collectors, yet far more than scale or arcade machine collectors. The interesting fact is that the population of gumbal, peanut, marble, breath mint, bulk candy and other glass globe and countertop vending machine collectors is growing faster than the numbers of any other coin-op collectible.

some countertop glass venders in the $25-$35 range at retail, an entry price level that no other vintage coin-ops can hope to enjoy. So collections can be built, and machines enjoyed, at reasonable investment levels.

That's not to say that these glass-globed beauties don't have their rarities. They sure do! A recently discovered Stanley HOT NUT peanut machine of circa 1925 sold a few years back for $1,750, and an offer of $25,000 was made on a Berger PEANUT VENDER of 1904, and turned down by its owner. Cast iron Columbus, Simpson and Advance machines of the teens, 1920s and early 1930s constantly command retail prices in the $200 to $500 range depending on the model, and all sorts of other globe venders that have never been cataloged or previously identified turn up all the time.

Belk, Schafer HUMMER, 1896. Vending machines started out as cigar sellers, adding toilet paper and water closet locks as technology developed. Hundreds of different cigar venders were produced, yet only a small number have been found. The Belk, Schafer PATENT CIGAR SELLER sold 10¢ cigars at left and 5¢ cigars on the right. Its name was changed to HUMMER once it became nationally popular. Maker was Belk, Schafer & Company in Alton, Illinois. Every 10th cigar was free. $3,720.

There are reasons for this. First, the expansion of coin-ops as a collectible is attracting new collectors. You can see more vending machines in antique shops, at shows, in malls or at outdoor fleas than any other class of coin machine because more of them were made. There was a time in recent history that every self reliant store, restaurant, cigar counter, hotel or motel lobby, YMCA counter, bus stop, train station or airport had one or more gum, candy or nut machines on the counter. Multiply that by the business establishments in the country in, say, 1939, and you'll be looking at machine numbers that no slots, jukeboxes, pins or anything else could ever match. The countertop venders were (and still are; look at the banks of them in the entryways of supermarkets and airports to this day) the highest population machines ever made. So they survive in larger numbers.

The other reason is their price. Most new collectors in any field don't start out with deep pockets. So they need to be able to buy and collect something that won't put them in the poorhouse. Not at first, anyway. And that's vending. Even the most sophisticated coin machine histories and price guides still list

Historically, the genre started in the late 1880s, although the early machines were generally wooden boxes with a mirror with the chocolate, stick gum or other vendables inside in columns. The glass globe models really began in the late 1890s with saloon placement to sell breath mints (to remove the odor of beer or whiskey, which they didn't do very well) and peanuts (to get that salty taste in the mouth to make the eater want more beer), with the contents stored in inverted glass jars over complicated cast iron coin-released bases on legs. By the early 1900s the globes had become dedicated design pieces with various machines having their own characteristic globe tops. Bulk mints and nuts were the basic vended commodities, which means you got more than one for your penny or nickel, although one-at-a-time vending of larger items soon became the norm. It wasn't until around 1907 that the Gravity Vending Machine Company of Chicago converted one of their boxy 4-column HONEY BREATH BALL venders to a one-for-a-penny gum ball vender, and the gumbal machine was born.

There is some debate whether the Gravity is the first gumbal machine (the confection gum ball itself came out earlier, a few years before the machine) but that's what makes vending machine collecting so exciting. It is really just in its infancy. While the odds are that the machine you have or find is usually a fairly common type, there is always that possibility that one of the rarities has reared its head. For that reason it is smart to accept the adage that "knowledge is power" and learn more about your machines before you buy or sell, trade or trash.

What you will be looking for are the many venders and stamp machines and peanut dispensers and other machines that were on the counters in "Grandpa's Store," or elsewhere (and were too nice or interesting to throw out!). As the antique market unfolds, these machines will be coming out of the woodwork to be identified and priced.

A word about "Grandpa's Store." Some years back, when I first started to collect "commercial interior" photographs, I found myself at a dead end. It was a very rare sight to find a photo of an old store, inside or out, that had any form of coin machine in it. I searched for the pictures. And realized that there had to be a better way to find them. That's when I hit on the concept of "Grandpa's Store." If the old coin machines were commercial devices, they had to be in commercial installations. And where would you find pictures of that nature? In family albums, and at local farm and family auctions, no less. Something had to pay to keep the family in Sunday chicken and nice clothes. It was THE STORE. I quickly learned that in any given family auction or album, if you were lucky one out of ten or more would show where the money came from, be it a store, coal yard, factory or whatever. At that point I started to get lucky and find photos and machines. When the two go together, they create a historical and monetary value greater than it's pieces. If you do have an old vending machine around the house, where did it come from? And is there a picture that shows it in use? Put those two things together and you have immeasurably increased the value of both.

In the teens and twenties and thirties of this century many of the family breadwinners were independent businessmen or women. It wasn't a big money deal, and often family livings were scratched out of small communities that supported their store or other commercial operation. But that's where the coin machines come from. Some are common. Most are, in fact. Some are rare. A few, that is. But many are wonderful pieces of Americana (Only in the U.S.A. was there such a diversity of commercial coin-operated machines!) that should either remain in the family, get sold to the collector market or end up in a local museum. You might try that some time if you need a tax deduction and your town would like to recreate the days of the past. That is where the combination photo and vending machine would come together neatly.

Here is a starter list to give you an idea of what kinds of machines to look for.

   Bulk Vending (Peanuts, other nuts, candies, gum, toys, etc.)
   Cigar/Cigarette
   Drink (Coffee, soft drink, hot or cold)
   Food and Sundries
   Gum (Stick and package)
   Gumbal (Also toys and marbles)
   Liquid (Gasoline, cleaning fluid, etc.)
   Matches
   Newspaper/Magazine
   Package (Candy and sundries)
   Perfume
   Postcard
   Product/Merchandise
   Sanitary
   Service (Shoe Shining, shaving, insurance, etc.)
   Stamps/Stationary
   Sundries (Combs, condoms, sanitary napkins, handkerchiefs, etc.)
   Timer (Parking Meter, Storage, Toilet, etc.)

Then there's the age and operation to consider. Sometimes the way a machine is made may help date it. The mechanism often do. Here are a few pointers related to materials that might help identify and qualify some collectible coin-op venders:

   Aluminum (Later production, most likely after 1932)
   Cast Iron (The best! Circa 1905-1918)
   Glass (Hard to place. Production began before 1905 and endured until after 1945)
   Molded (If it's plastic, it's fifties! Or later)
   Porcelain (Circa 1910-1935)
   Steel (Circa 1927-1960)
   Wooden (If you are lucky, before 1910. But many were made after 1930)

Volkmann, Stollwerck CONFECTIONS, 1897. After milk chocolate was developed by the Swiss in the 1870s the Stollwerck Company in Cologne, Germany, made chocolate venders. To sell the American market they established Volkmann, Stollwerck & Company in New York City through an agent. The German venders were very elaborate and expensive, so to meet the enormous needs of North America a vending machine factory was set up in Brooklyn. Two models were CONFECTIONS and MINTS. $3,220.

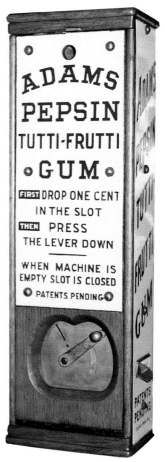

United States Slot ADAMS PEPSIN TUTTI-FRUTTI, 1897. The American (actually Mexican) contribution to the world's sweet tooth was chewing gum. By the 1890s gum makers, pioneered by Adams gum, led the way to a new national habit. Venders were on New York City "L" platforms and later the new subway. Maker is United States Slot Machine Company in New York, New York. Single column is "Little Adams," double column "Large Adams." Valuable machines. $2,810.

So much for conventional vending machines. The exciting part about collecting coin-ops is that things change. The coming and going of fads, and "phads" (meaning phenomenal fads), for instance. Soda machines come under the latter category. You can see it occur at the coin-op shows, and in the vintage coin machine media. Soda pop machines, essentially bottle venders, are hot. So hot, they have even "matured," with the genre settling down into a price realignment. Meaning we are finally beginning to know how many of one machine or another survived and can start to differentiate between rare, common, collectible and undesirable models. It is the classic bell curve of enthusiasm, analysis and acceptability, and something vintage coin-ops go through just as businesses and other enthusiasms.

The soda pop venders were first classified as "Coke machines," meaning Coca-Cola®, even though the collectibility went far beyond the red icon of American life. Little by little the collectors of these devices learned that there was life after Coke, and some enthusiasts for the machines became fond of saying "It doesn't have to be red." The meaning was apparent. There was also blue for Pepsi-Cola, mint green for Dr. Pepper, an off-orange for Royal Crown Cola, white for 7-Up and a few other oddball soda pops to round out the collectibility. That's the interesting thing about the soda machines, as they are now collectively called. More than any other form of vintage coin-op collectible and vending machine, they go by brand name. The makers, Mills Novelty in Chicago; Bevco in St. Louis; Cavalier in Chattanooga, Tennessee; the old Glascock in Muncie, Indiana; the massive Vendo of Kansas City (and now Fresno, California); the similarly named Vendorlator of Fresno (later bought out by Vendo); Westinghouse and others made their machines to order and sold them to the branded names on their cabinets. Westinghouse, for instance, sold only to Coca-Cola. Same for Cavalier. But others ran across the competitive gambit, with Vendorlator selling to all comers, such as Coke, Pepsi, RC, 7-Up, Dr. Pepper and others.

But deciding what's good, and what isn't so much of a find, requires knowledge as the variations are numerous. As in all fine antiques, and particularly in coin machines, you "gotta know the territory." That means if someone comes along and wants to sell you a very rare Pepsi-Cola Vendo V-39 from 1949 be wary. Vendo didn't make one, but a recent "restorer" probably did. The true soda pop enthusiast knows the difference. If you don't, you could get stuck.

Same for selling. With the values of soda pop machines fluctuating all over the map it would be to your advantage to have some idea what your box is worth before you attempt to sell it. A note of warning, because soda machine collecting is so new, and volatile, some of the machines that were barnburners last year aren't this year, and won't be next. Many have seen some dramatic price drops. Likewise, now that some of the rarities have been identified, other machines are increasing in value. It's like getting on on the ground floor of a new collectible, with all of the risks and possible gains.

Fleer TRY A SAMPLE, 1898. "Try one. You'll like it." If the product satisfies, or creates a want, the customer will likely come back. That was the theory behind the penny sample stick (Fleer called them tablets) of Fleer's Pepsin Gum TRY A SAMPLE machine placed in high density public locations. Frank H. Fleer & Company of Philadelphia made and owned their own gum machines. James Duke of The American Tobacco Company was doing the same thing with cigarettes. $2,610.

Hollands STAR PEPSIN, 1899. Gum became the most vended commodity in the country by the turn of the century, with the gum "Trust" fixing prices and owning the vending machines. It took years to break this hold on this popular new food fad. Almost every trust member made a pepsin gum, said to calm the stomach. Jet Hollands of Detroit, Michigan, patented STAR PEPSIN, which slid out two sticks for a penny. Production probably subcontracted to other Detroit makers. $2,010.

Freeport SODA MINT GUM, 1899. Freeport is a magic name in vintage vending machines. It means cast iron, elegance, distinctive mechanisms and rarity. Few Freeports have survived the years. The Freeport Novelty Company of Freeport, Illinois, made their own GOO GOO gum vender in 1899. They also made it for the Mills Novelty Company as GOO GOO, although the cabinet graphics are toned down. They used the same cabinet for their much less flashy SODA MINT GUM machine. $1,510.

Regina MUSIC BOX STYLE 17A GUM VENDER, 1900. Gum was on everybody's lips, so to speak. Late in 1899 long time music box maker Regina Music Box Company of Rahway, New Jersey, developed a vending attachment that coupled to their music box mechanism. Play the music and the clockwork mechanism of the music box pushes a stick of gum out from under a stack inside the machine and drops it into the cast iron dispensing lip at lower left. $7,510.

National Vending COLGAN'S GUM, 1902. The AUTOMATIC CLERK made clockwork action under glass a popular format, resulting in dozens of similar machines. The National Vending Machine Company COLGAN GUM, made in Chicago, gave you a stick of gum for a penny. Then every fifth penny got two. So you spent more to get it, or got a pleasant surprise when you used the machine. Taffy Tofu was a popular early 1900s gum flavor that went out of favor in a few years. Strange taste. $1,810.

Automatic Machine THE MERCHANT, 1901. One of the most sought after machines by vending collectors. Maker is the Automatic Machine & Tool Company of Chicago, the same maker of floor model slots and GABEL'S AUTOMATIC ENTERTAINER automatic phonograph. Sometimes misidentified as a Gabel THE MERCHANT as company founder John Gabel put his name on top of the cast iron cabinet. Cast iron peanut venders are among the most desirable of all vintage coin-op collectibles. $25,050.

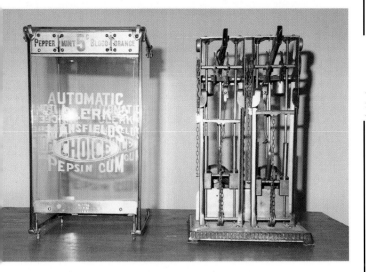

Automatic Clerk AUTOMATIC CLERK, 1901. A very long lived vending machine, used into the 1930s. AUTOMATIC CLERK holds two stacks of gum, choice of Peppermint or Blood Orange. A nickel in the desired chute starts the machine which automatically grinds its gears, elevates the chain drive and drops a stick your way off the top. Etched cabinet fits over the mechanism. You see the whole process. Made by Automatic Clerk & Advertising Company, New Haven, Connecticut. $610.

Winsted POTTER'S BEST, 1903. Before World War I much clothing, in fact most women's clothing, was made at home by seamstress maidens and sewing machine wives. Winsted Silk Company obtained the rights to a patented thread spool vender, calling it POTTER'S BEST SILK. Position the correct row with the color you want in the center behind the window by using the wheel at the top, put a nickel in the small cabinet below, turn the crank and out pops the spool. $2,510.

Pulver TOO CHOOS, 1903. Pulver wall machine development was rapid, with models changing almost yearly. TOO CHOOS is superficially refined over CHOCOLATE COCOA with a grille over the display window (people broke in and stole the dolls), much more modern graphics styling and less expensive trim. The inside character is "Foxy Grandpa," an early 1900s expression for a smart and with it senior citizen. Within a year Pulver machines got smaller, and this became the "Tall Cabinet." $1,210.

Price COLLAR BUTTON, 1904. Ultimately everything that could be vended usually was. Collar buttons were a constant necessity to keep celluloid collars in place on white dress shirts. Price Collar Button Machine Company machines, made in Iowa City, Iowa, were placed everywhere men went. Plain, square machines gave way to this colorful spinning carousel clockwork powered version in the 1904-1906 period. These machines keep showing up in old buildings. $1,510.

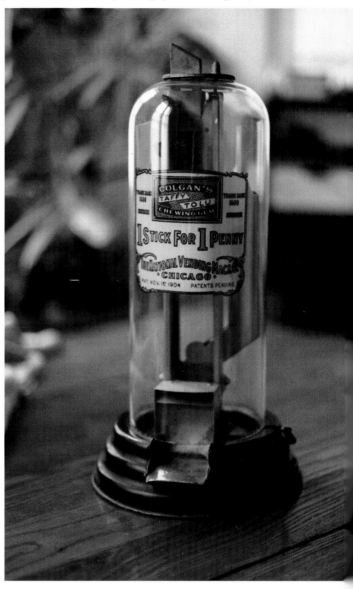

National Vending COLGAN'S TAFFY TOLU, 1905. Ultimate development of the visible action machines was the tall cylinder tower models. National Vending applied for a patent on their version in April 1905, and spun the machine off into three models: WILBUR'S CHOCOLATE, CHICLETS and the latest model of COLGAN'S TAFFY TOLU. This is the 1905 version, indicated by "Patent Pending" on it's label, on which their 1904 patent for match vending is mentioned. $1,460.

Austin, Nichols FOXY GRANDPA, 1905. Many cigar vending machines were dedicated to specific brands and generally identified the machine maker and marketer along with the cigar name. These panels are themselves collectible. FOXY GRANDPA (that name again!) placed by Austin, Nichols & Company, New York City, was a lawbreaker, but promoters often tried to get away with it. Government tax laws required cigars to be sold out of their box to prove payment of tobacco taxes. $2,810.

Automatic Sales IMPROVED TRUE BLUE, 1906. For a while there, glass cylinder tower model venders were seen everywhere. It was only when the glass started getting broken, and replacements were difficult to get (they still are!) did the machines start to disappear. Automatic Sales Company of Lansing, Michigan, made one for TRUE BLUE gum. Its plunger action didn't work well. So they improved it with an operating lever. $2,210.

International Vending SAFETY MATCHES, 1906. Cigar smoking required a lot of matches, so vending machines were developed for the need. Panels under the dome turn as matches are vended. International Vending Company of Chicago promised to make people rich by having them sell the space on the curved panels as advertising billboards. Few are found that way. Machine also found with the name Jackson Vending Company on the base. $1,210.

Gravity Vending GUM MACHINE, 1907. Believed to be the first gumball machine. Originally produced in 1903 as the HONEY BREATH BALLS machine by the Gravity Vending Machine Company (a play on the visibility of the mechanism in action) of Chicago, the mechanism and product tracks were modified to accommodate the new balls of gum just entering the confections market. GUM MACHINE resulted, with the product name "Gum Balls" used on the instruction panel for the first time. $2,210.

Advance CLIMAX CHICAGO, 1907. Advance Machine Company of Chicago spanned the Golden and growth years of counter vending. Formed in 1900, its machines survived well into the 1950s. CLIMAX, meaning the optimum model, was produced in 1901, patent issued in 1904. Improvements, including a wall rack, led to CLIMAX CHICAGO in 1907, with the Chicago name and later patent dates in the cast iron base castings. Some CLIMAX model is almost a must in any vending machine collection. $610.

Hilo Gum HILO GUM CLIMAX, 1907. It is difficult to keep track of the many models of the cast iron Advance Machine Company CLIMAX. When you add the versions made under private label for other companies you get utter confusion. Hilo Gum Company of Chicago provided vending machines at low prices or on lease, and then sold the gum to fill them. When they added peanuts to their offerings they needed a machine, so had Advance make a HILO GUM version of the CLIMAX. $910.

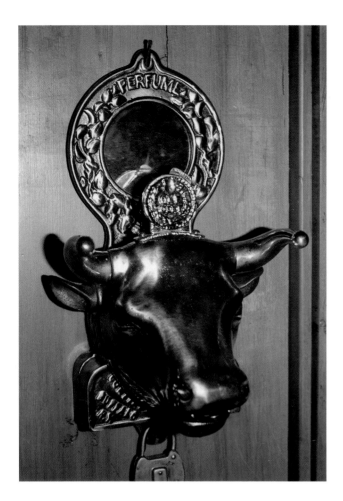

Continental Novelty PERFUME, 1907. A quick spray of perfume in the morning made you feel fresh all day. Or so the ladies thought in the early 1900s. They grabbed the cast iron bull by the horns, and swoosh! Got a shot. Actually, it was a bit strong. Made by the Continental Novelty Manufacturing Company in Buffalo, New York, who got the idea from neighboring Toronto, Canada, where the machine was patented in 1904. Its nickname is "Bull's Head Perfume." $2,610.

Jackson Supply THE HONEST CLERK, 1908. Were cigar countermen really that dishonest? THE HONEST CLERK made and marketed by Jackson Supply Company, Portland, Maine, gave you a good cigar for your money with no bait and switch. While this may have been a patented machine, no patent has ever been found. Maybe the "Pat Applied For" notation ended right there and they didn't get one. Cigar to be vended is in window, drops down at bottom. Boxful at top. $3,220.

Pope AUTOMATIC CIGAR SELLING MACHINE, 1908. The cigars are in their box with the lid open so you see the name and quality. You drop your nickel (or dime, depending on the model) and a mechanical arm reaches into the cigar box, selects a stogie, rotates halfway around and drops it in the receptacle in front. It all happens under glass as you watch. An inside moistener keeps the cigars mild and fresh. Maker and marketer is Pope Automatic Merchandising Company, Chicago. $3,020.

Clovena BREATH MINTS, 1908. As new store formations grew, and people got out more and bought things, every bit of counter space anywhere was put to good use. Breath mint venders showed up in drug stores and at the amusement parks where young people congregated, as well as in bars to cut beer breath. BREATH MINTS, made by the Clovena Machinery Company, has all the earmarks of a high density location placement, including an anti-slam arch over the plunger. $3,820.

Crystal CRYSTALETS, 1908. If a tall glass column looked great as a vender, two would look twice as great, right! The CRYSTALETS machine permitted display of two flavors of the hard candy and "Breathe Perfume." The instructions tell the customer to pull hard on the levers, with the candies dropped into the shared cups in the center between the two globes. Maker and marketer is Crystal Vending Company of Columbus, Ohio. $1,260.

Regina MUSIC BOX STYLE 18 GUM VENDING MACHINE, 1910. Regina changed with the times. The early STYLE 18 GUM VENDING MACHINE models of 1909 still carried the gum dispenser on the front at the lower left. But by the next year the cabinet had been restyled and the vender was unobtrusively moved to the right side, dropping a stick of gum from the stack inside when the music box was played. Regina called the STYLE 18 "A Musical Salesman." $7,520.

Hopkins & Kost FLEUR DE LIS, 1910. The novelty of automatic action under glass began to wear off as the machines wore out, or weren't wound up. The FLEUR DE LIS gum vender offered two flavors with a coin chute for each. Pick peppermint or wintergreen, drop a penny in the proper chute, and turn the wheel on the side. No more clockwork windups, or batteries. The vending energy was provided by the customer. Maker is Hopkins & Kost Company, Ltd., of Newark, New Jersey. $3,220.

173

Griswold PENNY PEANUT, 1910. M. O. Griswold & Company of Rock Island, Illinois, is said to have made the first peanut vender, presumably when the firm started in 1892 and for the World's Columbian Exposition in Chicago in 1893. None have been found. Later Griswold peanut machines, particularly those produced by the renamed Griswold Manufacturing Company, have shown up. PENNY PEANUT is a 6-legged cast iron model with an aluminum or brass receiving cup. $910.

Iowa Paper MATCHES, 1910. The trend toward customer activation grew rapidly, and by the second decade of the new century the glasswork automatics were becoming a thing of the past. The knob on the side became the hallmark of a new machine. Iowa Paper Company, of Waterloo, Iowa, a jobber of paper, woodenware and stationary, was a short lived producer of the new style venders, delivering a small box of their jobbed wooden matches for a penny. $810.

Advance GLOBE, 1911. Possibly the sole exception to the rule of the demise of the globe topped tall cylinder venders with working machinery inside. The GLOBE match vender was made by the Advance Machine Company in Chicago, starting in 1911. If you look at old store photos enough you will find the GLOBE time and time again. As long as boxed matches had a use, so did GLOBE. Advance cataloged it well into the '40s. The 1911 patent application wasn't approved until 1916. $610.

onal Novelty BREATH PERFUME, 1913. The major trend in small commodity vending, such [m]ints, peanuts and the now accepted gumball, was [to] rounded glass globe tops. The tall glass tops of [A]dvance and Griswold venders began to give [way] to the more conservative spherical look. Na-[tiona]l Novelty Company in Minneapolis, Minne-[sota,] a major coin-op scale maker, made BREATH [PER]FUME, although it was originally called the [TAB]LET VENDING MACHINE. $1,260.

Griffin Vending GRIFFIN, 1911. Perhaps the most audaciously beautiful cast iron peanut vending machine ever made. Griffin Vending Company survived in New York City for barely over a year, and then sank into oblivion. So did the machine. It was found in an antique shop late in 1992, having been made into a lamp. Its history was tracked down by the patent date on the back of the machine, and research in the business section of the New York Public Library. One known. $30,100.

Winters THE RED STAR, 1914. Eastern Iowa, along the Mississippi River, was a major producing and operating area for cigar and bulk product vending machines. Red Star Sanitary Vending Machine Company, operating out of Davenport, had RED STAR machines made by Griswold in Rock Island and H. E. Winters Specialty Company in town. Winters also made the machine for themselves as THE RED STAR. Two bulk product containers, left side 1¢ and right 5¢, suggesting peanuts and pistachios. $4,015.

Northwestern SELLEM, 1912. From a jump start in 1910, the Northwestern Novelty Company in Morris, Illinois, graduated from cigar specialties to become a major producer of glass globe vending machines. The firm's first coin machine vended a takeaway box of matches for a penny, while it held an open box in a holder for public use. Called the SELLEM MATCH VENDER, it became known as "Nobby." Similar machines were produced by a number of other makers. $360.

Lincoln (Simpson) LINCOLN HOT NUT, 1915. The classic period of round globe nut, confection and gumball machines began with the advent of the Columbus, Simpson, Hance, Ad Lee and other machines in the years just before World War I, most of which were made in and around Columbus, Ohio. The Lincoln LINCOLN HOT NUT was made by the R. D. Simpson Company, Columbus, Ohio, for the private label Lincoln Vending routes. Light bulb on top kept the nuts hot. $810.

Hance REX, 1923. Hance Manufacturing Company, located in Westerville, Ohio, just outside of Columbus, built private label for a number of the Columbus producers, particularly Norris Manufacturing Company. They also had their own line. Their REX breath pellet line evolved into this very flashy all aluminum polished body with toed feet. Two versions were made, differing in details. The breath pellets were great to mask the smell of bathtub gin. $1,015.

National Novelty THE NATIONAL, 1924. By the mid-'20s the gumball had become the primary vended confection, and there was a rush to produce enough machines to keep up with the surge of new business formations in stores, restaurants, roadside stands and other outlets related to the growth of cities and the explosive use of the automobile. National Novelty committed itself to the new business, changing its name to National Automatic Machine Company in 1927. $260.

Dietz SELF SERVICE, 1926. No blank trade stimulators gave you your money's worth, maybe more, over the counter. The SELF SERVICE vender gives it to you right away, automatically. Put coin in at the lower right, flip it with the lever below, and turn a crank on the side to get a package of gum. If the coin flips into the channel to hit Bozo, turn the crank again and get another. Maker is Andrew T. Deitz, Toledo, Ohio. Gum flavors are Rootberry, Raz Raz and Grape. $860.

Weeks AUTOMATIC PENCIL VENDER, 1925. Charles M. Weeks Company of Walden, New York, an arcade machine maker, created a vender that imprinted and personalized a pencil for a nickel. Front panel says "A pencil with your name 5 cents." Wooden cabinet version came out in 1923, replaced by this all metal model in 1925. A marketing company called Namon Pencil Company, Walden, New York, also sold it, as did Vendex, Inc, in New York City, both putting their names on the machines. $510.

Hance AUTO, 1928. Newer models after the REX breath pellet vender of the Hance Manufacturing Company of Westerville, Ohio, accepted the changes that took place with so many other globe venders; simplification. The Hance AUTO breath pellet machine has a simple base with a side crank to deliver the pellets. As typical of the genre, breath pellet machines are smaller than their peanut or gumball vending cousins, and are attractive pieces for small space collections. $710.

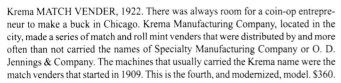

Krema MATCH VENDER, 1922. There was always room for a coin-op entrepreneur to make a buck in Chicago. Krema Manufacturing Company, located in the city, made a series of match and roll mint venders that were distributed by and more often than not carried the names of Specialty Manufacturing Company or O. D. Jennings & Company. The machines that usually carried the Krema name were the match venders that started in 1909. This is the fourth, and modernized, model. $360.

Stanley HOT NUT, 1925. Silent Sales Vending Company in Philadelphia, Pennsylvania, had R. D. Simpson Company in Columbus, Ohio, make a series of machines similar to the private label Lincoln hot nut venders with a shielded light bulb that warmed the nuts. The name "Stanley" was etched on the nameplate as shown here, or even included in the base as part of the casting. Later models had the Stanley name only in the paper on the globe. Rare. $1,770.

Ad-Lee E-Z BASEBALL, 1930. The baseball boom following radio broadcasts influenced vending. The Ad-Lee Company, Inc. of Chicago marketed machines produced by Columbus Vending Company of Columbus, Ohio, the largest producer of globe venders. The NEW E-Z model had a squat base and vended one drilled gumball for a nickel, with a paper message inside to make it a chance machine. E-Z-BASEBALL paid off on baseball messages, with the "Home Run" worth a dollar. $710.

Hoff WRIGLEY'S NO. 5A, 1931. In the 1920s the gum market was dominated by Wrigley's out of Chicago in a classic Snow White and the Seven Dwarfs business situation, with the next six in line a "fer piece away" from the leader. Vending machines were a major success factor. Hoff Vending Corporation in New York City made a line of Wrigley venders. WRIGLEY'S NO. 5A sold Juicy fruit, Spearmint, Double Mint and the smaller P-Ks. NO. 5B sold three stick gums. $360.

D. Robbins THE EMPIRE, 1932. So enormous were the routes and operator lists of D. Robbins & Company, located in Brooklyn, New York, the largest coin machine distributor on the east coast, they sought ways to reduce their expenses and maximize their profit. So they made their own multi-purpose vender. It was designed, and named, after the new Empire State Building in New York City, at the time the tallest in the world. Models handled peanuts, bulk candy, hot nuts, gumballs and toys. $460.

General Coin TYPE-O-METER, 1938. Get some work done in a railroad station, hotel lobby or men's club. It was possible with the TYPE-O-METER. 30 minutes typing time for a dime. A heavy clock spring is wound up when the coin slide goes in and starts counting when it comes back out. Marketed by General Coin Automatic Company, New York City and San Francisco. Made in Ilion, New York, with Remington STANDARD machines. Only one known, unless typewriter collectors have them. $1,210.

Rowe DELUXE, 1928. The major growth habit of the 1920s was cigarette smoking. After an early start in the 1890s, cigarette venders all but disappeared. Until the '20s, when WWI veterans who learned to smoke in service wanted easy access. Rowe Vending Machine Company in Los Angeles, California, was successful with their 1926 and 1927 STANDARD machines, followed by a DELUXE model in 1928. They became the leading producer. $460.

Columbus MODEL 34 GAMBLER, 1936. A nickel a pop for a candy ball seems excessive for the '30s. But that's because this is not an ordinary Columbus MODEL 34 penny vender. It's a gambling machine, along the lines of the Ad-Lee E-Z. The colored candy balls are numbered, and you check the colors and numbers against an award card on the counter to see if you were a winner of anywhere from a nickel to a dollar. Columbus Vending Company, Columbus, Ohio. $310.

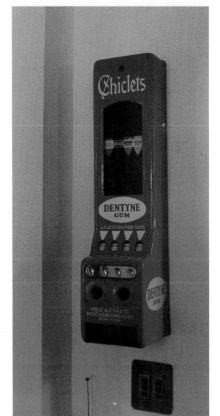

Mills Automatic MERCHANT, 1938. When Canteen Corporation stiffed Mills Novelty and pulled out their vending machine business, Mills formed the competitive Mills Automatic Merchandising Corporation, in Long Island City, New York. Special machines were created for the service and made by Mills Novelty in Chicago. MERCHANT vended various brands of gum. Some were sold to the massive Autosales Corporation, also in Long Island City, with routes all over the country. $260.

Industrial Carton CARTON COOLER, 1939. Successful vending of cold bottles of pop became a major achievement in the late '30s, with Coca Cola® leading the pack. CARTON COOLER is possibly the smallest coin operated cold vender ever produced, made by Industrial Carton Coolers of Lincoln, Nebraska. 8 iced bottles are held in a carousel, with 4 more cooling in the corners. Widely placed in Nebraska train stations and bus depots during WWII. 5¢ a bottle. $1,210.

Northwestern MODEL 33 PENNY BACK PEANUT, 1937. All of the glass globe vender makers had some form of chance model in their line. Both Advance in Chicago and Northwestern Corporation in Morris, Illinois, used a pinfield accessory that attached to the underside of a standard model, with the coin chuted through the machine onto the playfield. If it made it to one of the two return holes at the bottom you got your money back. Otherwise you paid for your peanuts. $560.

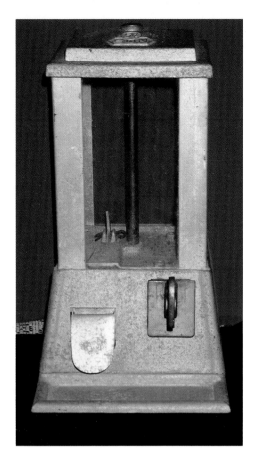

Los Angeles SUN, 1946. If any gumball vender can be said to be common, SUN is it. Due to its high population and low purchase price it is an entry level piece, and starts collections. Maker is Los Angeles Manufacturing Company of Los Angeles, California. SUN saw the light of day in 1946 as a nickel bulk vender of almonds, pecans and mixed nuts. Its flat glass and metal corners promised durability at low cost. Some have "5¢ SUN" on top; some don't. $40.

Silver King SILVER KING 1¢, 1946. Changing little over the years, a glass globe gumball vender called the SILVER KING was made by Automat Games in Chicago in 1938 and Automatic Games from 1939 until WWII, coming back in 1946 to be made by Silver King Corporation. 1¢ and 5¢ models were made, with the nickel version vending nuts. Production went well into the '50s. Other models included PRIZE-KING with two gumballs for each coin, and SUPERVENDER for oversize gumballs. $110.

Stoner NEW UNIVENDER MODEL 180, 1949. Pinball maker in the '30s, Stoner Corporation in Aurora, Illinois, converted to vending machines by the early '40s. Their UNIVENDER line of hiboy floor venders offered selections of candy, snacks and cigarettes. The unit was redesigned after the war as the NEW UNIVENDER. MODEL 180 is an 8-column candy bar vender. Coin chute at right of mirror takes 5¢ and 10¢, and makes nickel change when needed. $160.

Lawrence ROL-A-COIN, 1946. With World War II over, business bustled. It was before TV and fast air travel. A fast growing line was coin-operated radios for hotel rooms. Two hours of radio time for a quarter. ROL-A-COIN came plain, as shown, and in a DELUXE model that had a wake up alarm clock in the left panel and a program log in the right. Maker was Robert-Lawrence Electronics Corporation in Minneapolis, Minnesota. In a little over a year TV took over. $130.

Tropical Trading CHALLENGER, 1947. Offering three selections of hot nuts, the CHALLENGER was operated by the Tropical Trading Company of Chicago. The original CHALLENGER models had no nameplate on top, added later on. A left side cup vender was a standard addition to the machine. It was also produced as AJAX HOT NUTS with a nameplate at the top. Cebco Products Company used the machines as the TRIPLEX, with a similar two column model called DUPLEX. $260.

180

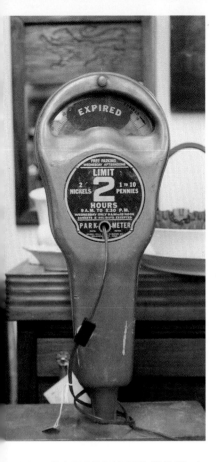

Magee-Hale PARK-O-METER, 1949. We take parking meters for granted. But they had to be invented. Carl C. Magee promoted their use, and formed the Dual Parking Meter Company in Oklahoma City, where meters were first installed in 1935. During WWII the Magee-Hale Parking Meter Company was formed to prepare for postwar sales. They came in a rush, with a lot of competition. The Magee-Hale PARK-O-METER was one of the first to be placed all over the country. $40.

A. H. DuGrenier ADAMS GUM, 1951. Pioneer in selective column vending of gum, candy, cigarettes and other goods, the firm has a convoluted history. Arthur H. DuGrenier, Inc. made venders in the 1930s and '40s, including a highly successful square corner ADAMS GUM counter and wall machine. The firm sold out after WWII to become DuGrenier, Inc. Employees bought out the new entity in 1948 and redesigned the lines. Among the offerings was a highly restyled 5-column ADAMS GUM machine. $70.

Coan U-SELECT-IT MODEL 74-AP, 1958. A major producer in the same league as DuGrenier, Coan went through a series of name changes in Madison, Wisconsin. Starting out as Madison Tool & Stamping, it became Rushour in 1929, and Coan-Sletteland in 1937. After WWII it came back as Coan Manufacturing Company. The redesigned U-SELECT-IT wall and console base venders were introduced in 1953. Models handle cigarettes, candy or snacks. MODEL 74-AP vends candy and snacks. $190.

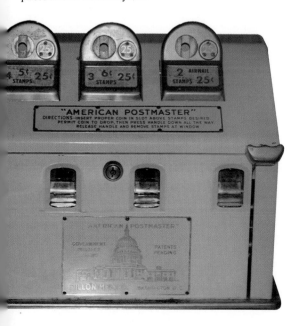

Dillon AMERICAN POSTMASTER, 1950. More machines have been made to vend postage stamps than any other type of vender with the sole exception of the early cigar machines, with stamp venders being made a lot longer. Strangely they have not caught on as collectibles and are greatly devalued. That will probably change. AMERICAN POSTMASTER was made in single, double and triple models. Stamp values changed with the rates. Dillon Manufacturing Company, Washington, District of Columbia. $110.

181

Stoner CANDY UNIVENDER, 1957. The hiboy floor venders of most manufacturers were so well worked out it was difficult to upgrade them. For a quarter of a century it was a matter of changing cosmetics. The Stoner CANDY UNIVENDER of 1957 has a larger mirror than the earlier models. An interesting addition is the gum or mint selector knob at the far left in place of the first of the eight columns. A small window shows the product. Many of these machines are still in use. $190.

Duncan DUNCAN METER MODEL 80, 1980. Parking meters have been with us for 60 years and show no signs of abating. A wide variety of cabinet designs and internal workings have been presented over the years. Local governments often sell off old models to help fund new ones. As yet no parking meter collections are known, but they would be a marvelous specialty interest. Duncan Industries of Elk Grove Village, Illinois, is a major supplier. This is MODEL 80 in use. $35.

Cavalier COCA COLA MODEL C-51, 1958. By the mid-'50s, over half a dozen producers were making soda machines. Cavalier made chest coolers for Coca Cola since the '30s, but it wasn't until MODEL C-51 that they were able to enter the floor model upright market. Holds and vends 51 bottles. Cavalier Corporation of Chattanooga, Tennessee, restricted their sales to Coca Cola while some makers produced machines for Pepsi, Dr. Pepper, 7-Up, Royal Crown and other bottlers. $610.

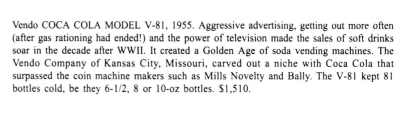

Vendo COCA COLA MODEL V-81, 1955. Aggressive advertising, getting out more often (after gas rationing had ended!) and the power of television made the sales of soft drinks soar in the decade after WWII. It created a Golden Age of soda vending machines. The Vendo Company of Kansas City, Missouri, carved out a niche with Coca Cola that surpassed the coin machine makers such as Mills Novelty and Bally. The V-81 kept 81 bottles cold, be they 6-1/2, 8 or 10-oz bottles. $1,510.

Coin-Operated Scales

If there is any area of vintage coin machine collecting that can be called new, it's scales. To be sure, there were always people who had scales, and there are even half a dozen dedicated coin-op scale collectors. But now a wider audience is opening up. One of the reasons for the lag is the fact that no books or articles have appeared in print touting coin-op scales, and what you are reading here is the first major effort in a book. This alone will probably start a collection or two. Another reason was the same reluctance that stalled pinball collecting; people didn't know you could own one of these commercial devices. They know now!

The latent desire for scale collecting goes back to an earlier era. Any reader over 40 years of age will probably remember them, standing on street corners, tucked inside movie theater and drug store doorways, often standing next to prescription counters so that patrons could weigh themselves and read the so-called "Health Charts" while they waited for their medicine and the drug store grabbed another penny. How many boys and girls growing up in the 1930s, '40s and '50s wondered if they were healthy or questioned if they would live past 20 years old when they read those weight charts that made everybody feel heavy and "unhealthy." It's an experience that present generations will never know (although you can recreate these troubled thoughts in the new solid state and digital scales in health stores to this day) and older generations choose to forget.

So the outdoor and indoor commercial scales went the way of the running board as society got tougher and broke the machines and people got more self conscious and didn't want anyone else seeing them weigh themselves in public. Weight went underground, surfacing only in diet programs and health clubs, and in our own bathrooms. So, through a combination of social and societal factors, the wonderful old coin-operated scales of the past disappeared.

Coin-op scales have an interesting history. The first one was widely recognized as the first popular coin machine. It's inventor was the Father of the British coin machine industry, an enigmatic gentleman named Percival Everitt. This incredibly creative London engineer invented amusement rides, bicycles, locks, date punches, postcard venders, grip testers and, in March 1884, a coin-operated scale. In his British patent application he stated that his "...invention consists in a new kind of weighing-machine constructed in such a manner that the act of placing a coin or the like in the apparatus will cause the weight of the body being weighed to be indicated on a dial or the like." Clever thing, that.

Everitt persuaded an old line British firm called George Salter and Company to produce it. He then trotted around the industrialized world patenting his device in Germany, France, Belgium, Italy, and throughout the British Commonwealth. Thereafter Everitt carted his big, cumbersome wooden cabinet round dial scale to the United States in March of 1885 to get it patented and produced by E. & T. Fairbanks And Company in St. Johnsbury, Vermont, by the end of the year. A coming recession quickly put Fairbanks out of the coin-operated scale business, but that modest beginning led to an enormous American scale industry, with dozens of models showing up wherever people congregated (and they congregated in droves with the new electric streetcars and interurban transit systems sprouting up all over the country) in the 1890s and early 1900s.

By the turn of the century enormous scales with very elaborate nickel plated cast iron cabinets were the vogue. Men, women, and children rushed to get weighed and watch the mechanism whirl and spin a pointer to indicate their weight on a large round dial. It became such a popular fad that songs were written about the phenomena. Images of Gibson Girls, showing these marvelous emancipated ladies on cast iron scales, were used in advertising, and a "Guessing Scale" (guess your weight within a pound and you got your penny back) grabbed saloon and cigar counter locations so that the beer swilling and tobacco chewing crowd could join the throng.

E. & T. Fairbanks WEIGHING MACHINE, 1885. The first public coin machine made in America. Invented by Britisher Percival Everitt, the "Father of Coin Machines." Patented in virtually all industrialized nations, plus some. He brought one to America for production by E. & T. Fairbanks & Company, St. Johnsbury, Vermont. One has survived the years, found in the basement of the local St. Johnsbury museum in the 1980s. Scale platform says "Everitt's Patent." $15,050.

The coming of World War I and the passing of Edwardian styling led to the "Lollipop" scale, a tall white porcelain coated cast iron machine that had an enormous "Big Head" dial that gave a wide viewing of the results when anyone stood on the scale. It was the beginning of leisure time and concern for ones' appearance. But the "Lollipop" had the seeds of its own destruction built-in. The flappers of the twenties, viewing flat bodies and skinny frames as their ideal of beauty, didn't want the whole world to see what they weighed, so by the end of the 1920s the Big Head scale was dead in the water. In its place came the ticket scale, where the weight was imprinted on a card (often with a horoscope as well as a movie star picture on the other side) while the scale still whirled and spun; only it was a fascinating spinner that gave the performance, not a giveaway pointer. The ticket scales added another fashion feature. As more women than men were now using scales, the designers added hooks to hold purses and shopping bags while the ladies weighed, or held a cane or an overcoat for the men.

It was the small waist high "Personal Scale" that captured the market of the 1930s, and has ever since. In a "personal" you look down at a wheel or a dial, and only you can see the results. It was about this time that the "Health Charts" were added to the purse, bag and coat hooks, making weighers wish they could meet the charted ideal, which kept them coming back for another weigh to see how they had progressed. The big names in the personals were Watling, Pace, Rock-Ola, Mills and others, with their scales surviving for many years. Pace went out of the business with America's entry in World War II, selling their inventory, tooling and production machines to other producers after the war, while Rock-Ola came back after the war for a few more years. It was Watling that survived, until the 1970s. Watling scales can still be seen in some out of the way places. In recent years some solid state scales have appeared that also count calories and provide a horoscope. But the day of the penny scale has gone, with the dime and quarter doing the job for today's operators.

The coin-operated scale is a handsome and enjoyable collectible because, of all coin machines, it more clearly invokes its period of greatest activity than any other. It does so because it was both an outside and an inside piece, and as such added to the ambience of its times. It hasn't changed that much. Arcade machines have gone video; jukeboxes are CD; modern pinballs are a truly kinetic experience enhanced by sound and speed; slot machines are solid state playable plethoras of light and visuals; trade stimulators and counter games are extinct and long gone; and vending machines look completely different than they did just a few years ago. But the vintage weighing machine goes on and on, a working piece surviving from a past age to the present. The penny scale was as much a piece of furniture as it was a commercial coin grabber. It was also a major part of period decor. If you see a photograph of an old Big Head or lollipop scale in front of a drug store your mind can almost recreate the look of the street, mentally placing old roadster rag top and boxy automobiles (they weren't really called cars yet) in the narrow lanes and parking spots. The same machine in an inside location invokes images of tiled floors, tin ceilings, marcelled hair, match venders, malted milks and all the other small details that made up the 1920s and the '30s.

The smaller "personals" recreate their own settings. Stuck in corners, or at the entrances and exits of stores (with the ever popular "Dime Store," the discount center of its day, a primary location) you can feel the 1930s and into the '40s by just looking at the machines. Often the style is Deco, or ultra Deco reaching into the area of Moderne, that stepchild of industrial design. Then, at the end of the '40s and well into the '50s, '60s and barely into the '70s, the Watling TOM THUMB look took solid hold, with a raft of clones made by Marion, American, Rx, Public Scale and others. With them you can relive the days of roadhouses, the first of the fast food hamburger joints (that McDonald's replaced) and the shopping malls, motels and strip store fronts that have characterized the American scene from the '50s forward. Even today you can still see older scales on locations in spite of the newer digitals and fortune tellers that are slowly but surely taking their place.

Percival Everitt LIEBES-WAAGE, 1889. "Everitt's Patent" coin operated scales were either produced or patented to be marketed in Great Britain (George Salter & Company, West Bromwich, England), Australia, India, New South Wales, New Zealand, Queensland, Tasmania, Victoria, Austria-Hungary, Belgium, France, Germany, Italy, Luxemburg and Spain, and should be found there. LIEBES-WAAGE, or "Lover's Scale," is the third German model. Scale platform says "Everitt's Patent." $15,050.

Chicago Recording AUTOMATIC RECORDING SCALE, 1899. Stand on the platform, drop in your nickel (expensive for the 1890s!) and you got a printed ticket with your weight while music played. It all happening under glass while the printer worked and the music box rolled, certainly your money's worth. And you got a souvenir of the new weighing habit. Made by Chicago Recording Scale Company in Waukegan, Illinois. This one was found in a New York City railroad station when it was razed. $16,050.

National Weighing NATIONAL, 1900. The money to be made in scales led to large operating companies that even sold stock. First large operation was the The National Weighing Machine Company of New York City. They made, subcontracted or bought their machines from 1887 through 1902. Best known was their cast iron NATIONAL made in a variety of versions, all with similar cabinets. Marquees were plain castings or in blue or red porcelain. This is the NATIONAL "Correct Weight" model. $1,810.

National Automatic NATIONAL, 1903. It looks the same as the National Weighing NATIONAL scale of 1900, and it is. Except the name has changed. National Weighing Machine Company scales had their name cast into the step plate on the platform, and included in the porcelain face dial. When the operating company was reorganized as the National Automatic Weighing Machine Company, still located in New York City, the face and step plates were changed to the new company name. $1,810.

Watling STYLE 1 GUESSING SCALE, 1902. Primarily a floor model slot machine producer, the Watling Manufacturing Company in Chicago added a chance device scale to their line. It was a unique idea, and took off. Get on the scale, set a pointer to your assumed weight by means of a knob on front, drop in a penny and wait. If you are within a pound of your weight you get the penny back. These became known as "Penny Back Scales" and "Guessers." Very desirable piece. $4,520.

National Novelty NATIONAL WEIGHER, 1899. In the beginning coin-op scale manufacturing was dispersed all over the country as entrepreneurial producers tried their hand at the new business opportunity. Minneapolis, Minnesota, soon had a number of makers. The leader, and surviving firm, was National Novelty Company, also an arcade machine producer, who made scales into the 1930s. This cast iron cabinet model was called "Round Column" based on its style. $2,010.

But the advanced machines have a long way to go to replace the mechanical devices that can be found on routes all over the country and throughout the world. Take a train trip through Europe if you want to see older scales grabbing coins as they have for the past century. You'll find them everywhere, in the U.K. and all of Europe, with probably higher populations than you will find in the United States.

Somehow through the years the scale has endured, offering utility, style and history. The coin-op scale does exactly the same thing in the same way now that it did when it was first introduced in the 1880s, over a hundred years ago. True, the mechanisms may differ (although the original British Everitts' mechanism of 1884 is still in use in Salter scales on location in Australia and Southern California as operated by an enterprising entrepreneur, a claim that no other form of coin machine can make) and their appearances might have changed. But not that much. Their form still follows their function, and people still drop a coin in machines that tell them what they weigh, sometimes getting even more information for the money or for an extra coin if called for. And for those functions a vintage machine as often as not serves just as well as a modern one. A major scale operator and antique scale collector named Bill Berning places scales on a national, and even international, basis, using vintage Watling scales and others as a matter of course.

As collectibles, scales add grace and dominance to a room, depending on the style and format. A small "personal" in the bathroom is considered quite stylish, and a massive lollipop in a den or kitchen demands attention while it adds a semblance of the industrial look to an active work area. But their attractiveness goes even beyond their good looks, for scales are usually quite easy to maintain. Except for a few rare cases, scale mechanisms are quite simple and easily readjusted once they get out of whack. As long as all the parts are there, and the castings or cabinets haven't been cracked and warped out of kilter, a coin-op scale can be depended on to give service as well today as the day it was produced. If you are lucky enough to have the original working manuals with your scale you are a leg up on maintenance. But even if you don't, chances are you can obtain this information. Most scales were made by the major scale firms. One of the most common collectibles is the American Scale Model 530 COMBINATION SCALE that gave you your weight and a fortune for 5¢, or weight alone for a penny. It was introduced in the summer of 1955 and was widely sold and operated. Soon thereafter, in the sixties, scales suffered from public vandalism and destruction, and many were broken into or destroyed. Only now, a quarter of a century later, are they coming back. Most "personal" (no big dial; only you can read it) chest-high scales from the 1946-1966 period are valued in the $150-$450 range depending on model and condition, with some Watling models going higher. The American Scale models are down the line after Watling, Pace, Specialty and some of the earlier more Deco looking American Scale models, with the red top Model 530 and green top Model 531 selling around $125 as collectibles.

Are there any old scales still out there? Strangely, after all these years they are still being found. Some surfaced when the New York Central Station was razed in New York City. They were scales from the late 1890s, buried in storage rooms. A stash of old Pace and Rock-Ola scales showed up in Chicago in 1992 still in warehouse storage. And every once in a while you can see one still taking in the coin, warning children of the future that if they are 5'6" tall they should weigh such and such. And it's still frightening.

As for maintaining a collection of scales, vintage scale operator and collector Bill Berning is dedicated to keeping the old machines in operating condition, and will gladly send free parts diagrams and maintenance drawings for Watling, Mills or Pace scales for a pre-stamped self addressed envelope. Write to: Bill Berning, P.O. Box 41414, Chicago, IL 60641-0414, and specify the type of scale you have. Bill also has reproduction parts for most scales to get them back into operating condition with their original good looks.

Watling STYLE 4 WEIGHING MACHINE, 1904. When the "Penny Back" Watling STYLE 1 GUESSING SCALE turned out to be so popular, they found themselves in the scale business. Taking the basic cabinet and mechanism, and removing the payback guessing feature, Watling had an attractive "Plain Vanilla" model. The money back cup was removed and the hole covered, and a new glass without all of the special instructions was substituted. Desirable, but not as much as a guesser. $2,510.

Caille Bros. AUTOMATIC BEAM SCALE, 1904. With Watling in the scale picture, it wasn't long before coin machine makers Caille Brothers Company in Detroit and the Mills Novelty Company in Chicago joined them in the business. The typical doctor's beam scale seems a strange device to put under coin control, yet it provided the most accurate and trustworthy weight possible at a time when spring scales were suspect. Inserting a coin releases the beams, which you then set manually. $2,410.

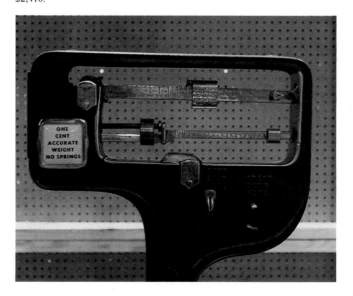

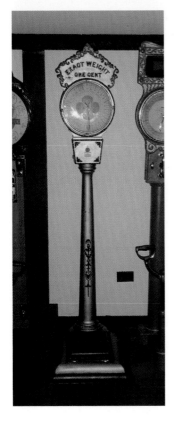

Mills Novelty WEIGHING MACHINE, 1904. The evolution toward the "Big Head," or "Lollipop," scale can be seen in the WEIGHING MACHINE made by the Mills Novelty Company in Chicago. The dial isn't big yet, but it's a start. Mills saw the scale as a necessary addition to their coin machine lines and enormous distribution potential. Soon the country was flooded with them. In terms of design it owes much to the National Novelty NATIONAL WEIGHER "Round Column" of 1899. $1,610.

Caille Bros. WASHINGTON, 1905. One of the top and most desirable coin-op scale collectibles because of its terrific graphics, fine wood cabinet, incredible nickel plated trim, and the fact it was made by Caille. The very colorful printed dial face identifies this as "The Washington Scale." A "Penny Back" model was also made as the WASHINGTON MONEY BACK. Among collectors these are known as the "George Washington" scales. A shorter WASHINGTON CUT DOWN model was made in 1910. $8,520.

United Vending WEATHERPROOF TALKING SCALE, 1908. A scale that speaks your weight. In 1908! The idea goes back to the British in 1887. This version was created by the Moore Talking Scale Company in Boston in 1902, becoming American Talking Scale Company of New York City. It didn't work well. Rights were then sold to the United Vending Machine Company in Cleveland, Ohio, who re-engineered it. A needle hits on a 10" disc record to speak your weight. Rare and valuable. $9,015.

Mutual Automatic MUTUAL HONEST WEIGHT, 1908. Copycatting, or private label imprinting, became endemic to coin operated scales once the enduring fad for weighing had spread across the country. Mutual Automatic Machine Company of Omaha, Nebraska, introduced a square column cast iron scale that was to all intents and purposes a duplicate of the National Novelty Company NATIONAL WEIGHER "New Open Dial" scale of 1906. National Novelty may have made it. $2,260.

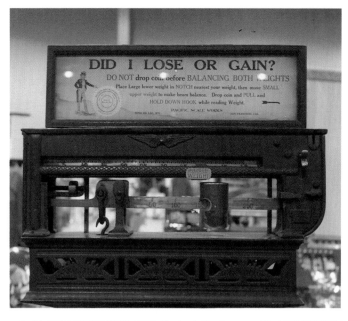

Pacific PUBLIC WEIGHER, 1912. Charles A. Fey, inventor of the LIBERTY BELL slot machine, had close personal ties to a friend in Fond du Lac, Wisconsin. So he spent a lot of time there. He patented a coin-op beam scale, and had it produced by the Pacific Scale Works in Fond du Lac. Paper on the scale notes offices in Wisconsin and San Francisco, California. This cast iron scale is attractive and complicated, and greatly enhanced by its instruction paper. $2,010.

Watling Scale MODEL 5 ALL STEEL CABINET PLAIN WEIGHER, 1916. There were two Watling scale companies. The first, Watling Manufacturing Company, headed by Tom Watling, made slot machines and scales. When older brother John, a partner, had a falling out with Tom he formed his own Watling Scale Company to just make scales, also in Chicago. He even got his own money back scale patent. His steel cabinet model was made both ways, "Penny Back" and plain. This is the plain. $1,210.

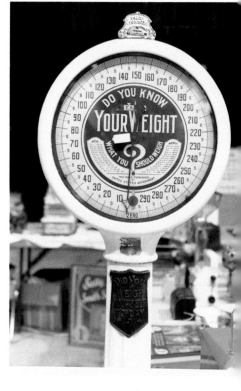

Autosales AUTOSCALE, 1928. In the late 1920s the old NATIONAL scales of the National Automatic Weighing Machine Company were up for another round of revamping to upgrade their looks. Autosales Corporation had been formed in New York City as yet another scale operating entity, selling stock. Existing NATIONAL scales were given a new steel sheath around their cast columns, plus a new green and red color scheme and an Autosales logo decal. Their new name was AUTOSCALE "New National." $2,010.

Mills Novelty 1918 ACCURATE SCALE, 1917. Mills Novelty was smart. They called their scale the ACCURATE, suggesting it was. But as a spring scale, it gets a little goosey after a while. The main springs constantly need replacement. Mills adopted the "Big Head" white porcelain clean look already claimed by Watling in Chicago and Caille Brothers in Detroit, and it became the style to beat or meet. It remained the industry choice for a dozen years. Mills made the 1918 ACCURATE for ten. $1,610.

Automatic Beam AUTO SCALE, 1911. The "Big Head" scale is on its way. The Automatic Beam Scale Company of New York City entered the scale business with a series of coin operated beam scales. By 1911 they gave that up and went for the growing popularity of round face scales. The style closely follows that of the earlier NATIONAL scales, with a much lighter round column cabinet. Cast iron was well liked as its weight deterred scale theft, as most machines were left outdoors. $2,210.

Peerless ARISTOCRAT, 1919. When Caille Brothers created their ARISTOCRAT scale in 1916 they joined the trend toward lollipop scales. The revolutionary design actually came out of a Watling model and a series of health scales made by the Toledo Scale Company in Toledo, Ohio. Caille helped commercialize it. When Caille Bros. set up the Peerless Weighing Machine Company in Detroit to compete against the Autosales Corporation routes, they put ARISTOCRAT on their routes. Peerless models came in white, red and blue porcelain. $1,260.

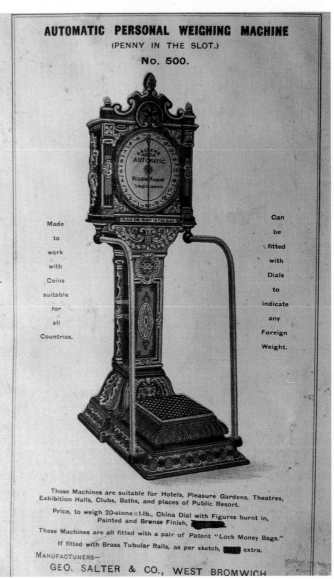

Salter NO. 500 AUTOMATIC PERSONAL WEIGHING MACHINE, 1906. The British approach toward coin-op scale development was far more elegant than the pragmatic American attitude, with massive size and color contributing to the entertainment of the machine. The whole advanced world was taken up in the fad of weighing in public in the early 1900s. It is not known if any George Salter & Company NO. 500 scales have survived. It would be a tremendous collectible. $210.

Peerless ARISTOCRAT DELUXE, 1921. The hallmark of a Peerless machine was its front shield, usually saying "Did you weight yourself to-day?," the idea being to generate a penny sale in the name of health. ARISTOCRAT DELUXE is a refined model of the original 1916 Caille Bros. ARISTOCRAT, with a completely new mechanism under the platform. As a result the ARISTOCRAT DELUXE base is thinner than that of the ARISTOCRAT. It is one of the most beautiful coin-op scales ever made. $2,510.

National Novelty MODEL 103 NATIONAL, 1922. After years of cast iron pole scales, National Novelty in Minneapolis finally went for the lollipop style. They did it in a beautiful white porcelain scale that looks somewhat like the Caille Bros. ARISTOCRAT DELUXE, except the perimeter of the circular big head is composed of twelve flat surfaces. The appearance is distinctive. Platform step plate is bronze, with the company name in the casting. Called "New National." $1,610.

Columbia Weighing COLUMBIA MIRROR, 1922. Developed in parallel with the big head scales, cabinet machines were simpler to produce and were less expensive. Columbia Weighing Machine Company, Inc. made scales since 1902, and was currently located in the Bronx, New York. COLUMBIA MIRROR has a wooden cabinet and was built for their own routes. It is usually found in the east. The mirror provides confirmation of your weight, or loss. COLUMBIA MIRROR CHIMES model was also made. $1,210.

Columbia Weighing HONEST WEIGHT, 1928. Columbia Weighing eventually produced a lollipop. Like others, it appeared to be almost a clone of the Caille Bros. ARISTOCRAT DELUXE, although it has its own unique dial. Production lasted for some time as evidenced by the step plate. Around 1930 the company changed its name and address again, becoming the Columbia Scale Company, Inc. in New York City. It's on the step plate casting. The Columbia Weighing name is on the face. $1,265.

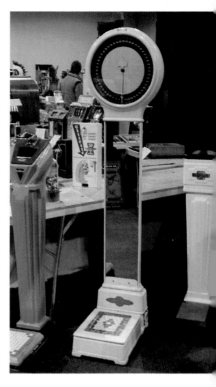

Peerless ARISTOCRAT JUNIOR, 1924. Peerless Weighing Machine Company had something going with the Aristocrat name, which ended up being used on all sorts of scales that had no connection with the original. The popularity of models such as the COLUMBIA MIRROR led to a mirrored front scale called the ARISTOCRAT JUNIOR, produced solely for Peerless use and not made as a Caille model. Its nickname was "Mirror." A plain metal front version was also made. $910.

Watling GAMBLER, 1929. The premier maker of lollipop scales, Watling Manufacturing Company even had a two story picture of one on the side of their Chicago factory. Watling big heads are big, and were made in a profusion of models. GAMBLER is unique as you have the option of picking one of three slots, not knowing which one will return your coin. It changes. Watling stresses its superior no springs design in its big head glass. Beautiful. $2,210.

National Automatic MODEL 170 NORMANDY CHIMES, 1929. Probably the most sought after cabinet scale, the National Automatic Machines Company NORMANDY CHIMES, made in St. Paul, Minnesota, has a money back feature. If it goes into the proper channel you get the penny back. If not it plays the chimes as the coin works its way down the cabinet, hitting bells in full view along the way. You see your weight so you always get value. And even if you lose you get those chimes. $1,810.

International Ticket MODEL A INTERNATIONAL TICKET SCALE, 1929. Ticket scales finally got away from showing your weight off to the world by printing your weight on a small card, with a fortune on the other side. To keep your attention the geometric spinner in the center whirls around in a psychedelic blur to let you know the machine is working. Then, plop! The printed ticket drops into the cup. Made by International Ticket Scale Corporation, New York City. $760.

Public Scale WEIGHT VERIFIER, 1929. Just as Columbia Weighing and Columbia Scale made machines almost exclusively for New York and eastern routes, Public Scale Company in Chicago made them for Chicago and points west. Styles followed current practice. But in 1929 they reached back to a design created by the earlier People's Scale Company in Chicago in 1914 known as the TRIPLE PILLAR, only they stuck a lollipop head on it. It too became known as "Triple Pillar." $3,215.

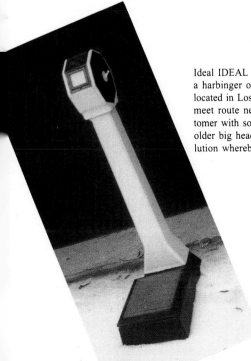

Ideal IDEAL SCALE, 1929. Even in the 1920s southern California was a harbinger of things to come. Ideal Weighing Machine Company was located in Los Angeles, California where their business was operating. To meet route needs they developed the IDEAL, both to provide the customer with something new as well as reduce the excessive weight of the older big head machines. It contributed to a personal scale design revolution whereby only the person weighing can see the results. $235.

Barnes NAVCO, 1930. Coin machines have long been a rapid reflection of popular culture and public taste, and remain so. The marvel is that changes in attitude are so clearly defined. When the personal scale of National Vending Machine Corporation in Detroit made its bow in 1930 it was a Deco delight, and light years ahead of any competition. The machine was built by Barnes Scale Company of Detroit, Michigan, and named NAVCO after the operating company. Tiled platform. $310.

Caille Motor MODERNE, 1930. The original Caille Brothers Company wasn't doing so well in slot machines and at the beginning of the 1930s had even changed its name to the Caille Motor Company in honor of its outboard motor line. Perhaps the last Caille coin machine product under the old aegis was the MODERNE personal scale. It owes a lot to the NAVCO in design, yet it is different. Part of the allure of this machine is its wonderful selection of early '30s colors. This is the black and cream model. $310.

Colonial FAIR WEIGH GOLF, 1930. Novelty scales are in great demand. FAIR WEIGH GOLF, made by Colonial Scale Company of Boston, Massachusetts, is perhaps the greatest novelty scale of all. It works somewhat like the NORMANDY CHIMES in that the penny works its way down the front of the machine under glass. Only this time it traverses a simulated golf course. If it makes it through the obstacles to the bottom you get it back. Dial is discreet, and windowed. Two known. Valuable. $6,520.

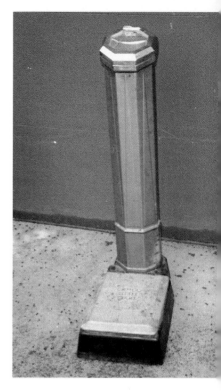

Camco scale LITTLE GIANT, 1931. An enormous manufacturing and operating combine was created in 1928 with the formation of Consolidated Automatic Merchandising Corporation of America, or C.A.M.CO., picking up the Peerless scale routes in the process. The entity became known as Camco, with a scale producing division renamed Camco Scale Corporation. It developed the LITTLE GIANT personal scale, going to the Peerless routes. The LITTLE GIANT was in use into the '40s. $410.

Mills Novelty MODERN SCALE, 1931. Mills Novelty always went for simple names. MODERN SCALE. That says it. And its Deco just about beat the pants off all comers. It was offered in six color combinations, so a collection of the six variants of the Mills MODERN SCALE would be a colorful thing indeed. The mechanism is very small, leaving the entire hollow cast column as a cash box. Find one full of 1930s coins and you'll have something. $310.

Mills Novelty MODERN SCALE WITH HEALTH CHART, 1931. Later in the year Mills brought back the MODERN with a large marquee at the top. Health charts! If you are so tall you should weigh such and such. And nobody ever did. So Mills kept getting people to weigh themselves to see if they were getting closer. Small cast aluminum models were used by Mills "Roadmen," their name for salesmen, to put the machine over. Today this model is worth more than one of the scales. $260.

192

ock-Ola IMPROVED LOBOY MODEL B, 932. Rock-Ola started making scales in 1927. hey brought out the FEATURISTIC personal ale in 1930, improved it as the LOBOY, and en made improvements in the LOBOY. They so made the scale private label, so you will d them with all sorts of other company names the platform step plate casting, or the name- ate on the front. These scales are very, very avy as they are thick porcelain coated cast n. I mean heavy! $260.

International Devices ARISTOCRAT, 1933. Watling, Rock-Ola, Mills, Caille Motor, smaller firms, and the Pace Manufacturing Company in Chicago all made new personal scales, and they were all selling. Pace made an attractive model called BANTAM, the same name used for a slot machine line. It was also made private label for resale. One buyer was International Devices Corporation of Chicago, promoting the scale as the ARISTOCRAT. The name is in the step plate casting. $260.

Watling IMPROVED TOM THUMB, 1935. The Watling contribution to the new personals was a small and uniquely designed scale called the TOM THUMB, using a big head mechanism packed into a mirror front cabinet. It came out in 1930 and was continually improved from year to year, surviving in one form or another until the 1970s. The IMPROVED TOM THUMB made use of a new mechanism specifically designed for the machine. These are well liked. $310.

A. S. L. Sales ACCURATE SCALE, 1936. Not all personal scales are good looking. Some are unattractive, if not downright ugly. The A. S. L. Sales Company ACCURATE SCALE, made in Dayton, Ohio, has a chrome top on a porcelain column that squeezes in the male and female health charts on each side of a drum dial. It looks somewhat like a bug's eye. But at $35 the ACCURATE SCALE sold like popcorn. Its common name was "Penny Scale." Available in white or green with black trim. $130.

American Scale FORTUNE MODEL 300, 1937. American Scale Manufacturing Company located in Washington, District of Columbia, grew slowly but surely to become the major competitor to Watling. But that was far away when the FORTUNE MODEL 300 came out. Standard if not dull in appearance, MODEL 300 carries its health charts baked into the porcelain on front. The feature is its four coin slots. Pick the right one and you get the penny back. If not, it rings a chime. Fortune dial. $260.

Raizner ADVERTISING SCALE, 1937. Scale revamping was almost as prevalent as slot machine rebuilding, and often not as subtle. Starting with a Watling IMPROVED TOM THUMB, a New York route operator named Nathan Raizner modified them to add a neon sign in a new topbox. The idea was to advertise the location where they were placed. This version lights up "Save At Mishkin," a local sundries store. Raizner got the idea patented. Only a few of these have been found. $660.

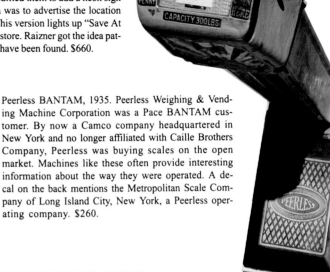

Peerless BANTAM, 1935. Peerless Weighing & Vending Machine Corporation was a Pace BANTAM customer. By now a Camco company headquartered in New York and no longer affiliated with Caille Brothers Company, Peerless was buying scales on the open market. Machines like these often provide interesting information about the way they were operated. A decal on the back mentions the Metropolitan Scale Company of Long Island City, New York, a Peerless operating company. $260.

Kirk GUESS-ER, 1938. In an era of personal scales, C. R. Kirk & Company in Chicago flew against conventional wisdom and produced a large dial guessing scale. It was the same idea Watling had used years before; set the dial with the front knob, drop the coin, and if you were within a pound you got the coin back. Only now it was electrically lit up and the venture cost a nickel. Also made for export, with this one found in Australia. $810.

Public Scale GUESS, 1938. It seems everybody was doing it. Chicago's Public Scale Company was matching every scale by every maker to be sure they were available for Chicago routes, a franchise they held through political clout (a wonderful Chicago word). GUESS worked the same way as Watling and Kirk; set the dial, drop the coin, get your money back if you were within a pound. The difference is one of mechanism and cabinetry. $610.

Public Scale PERSONAL SCALE, 1939. In matching everyone else's scales, Public Scale Company also made a PERSONAL SCALE. In fact, they made two, one quite complicated and another with the same mirrored column and mechanism with a simpler head, as shown here. It was the mechanism that dictated the design, for it was the same one they used on their large scales. That made the Public Scale PERSONAL SCALE large for its genre. Rare, but not particularly desirable. $165.

American Scale MODEL 403, 1950. The scale business nearly disappeared in the '50s. Malls, interstate highways, fewer truck stops, all that. Plus just about everybody finally had a scale at home, leading to a greatly diminished potential for coin-ops. MODEL 403 dropped the penny back feature, and had 12 slots on top by months. Pick your birth month and get a reading on the drum reel. The penny return cup is covered with a plate that says "In what month were you born?" $135.

Watling 1948 FORTUNE TELLING MODEL 400, 1948. After the war, Watling survived as the Cadillac of personal scales. The most popular Watling was the fortune teller, because you ask the question (among 21 knob set possibilities) and the machine answers on a small drum dial window forward of the weight window. The MODEL 400 kicked that feature up by offering 200 different fortunes, and said so on a mirrored marquee. Very desirable as it is fun. $460.

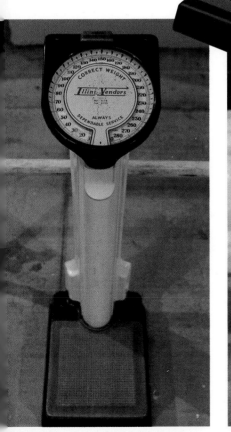

Hanson PENNY SCALE, 1939. Hanson Scale Company was a Chicago maker of coin counting scales, and a round dial "Health Scale" for the well to do and athletic clubs, forerunner of the modern health and exercise club. It didn't take much to add coin-op. Made from 1930 to 1939, when it was redesigned with a thinner platform so ladies in high heels wouldn't trip over it as they did on the older models. Dial could be personalized by operator; this one is Illini-Vendors. $160.

American Scale MODEL 400, 1947. Right after World War II the demand for scales was as great as it ever was. That put American Scale Manufacturing Company in a good spot for postwar business. Slab sided steel, and simple to produce, the Americans undersold a lot of the market and gained machine population. MODEL 400 kept the four coin chute of its prewar models, so you got your penny back if you guessed right. Plus your fortune in a drum dial. And weight. $110.

Watling BIG GUESSING SCALE, 1941. When John Watling died, brother Tom took over his Watling Scale Company business, adding some big machines to the successful TOM THUMB line. One floor cabinet model offered enough room for a new variation on the guesser scale, making it almost a personal scale in concept. Knob set the left drum dial to your weight, and when you pay the right drum dial spins into place. If you're close, the penny comes back. And only you can read them. $1,615.

American Scale COMBINATION MODEL 530, 1968. Down and dirty, slab sided and cheap, the American Scale COMBINATION MODEL 530 and its variants took a penny on a side chute for your weight or a nickel in your birth month in the round marquee on top for your weight plus a horoscope reading. They stressed the machine didn't need constant scroll or capsule refills, like Watling. 530 has a red top; 531 green. "M" models have mirrored marquees. Lowest value collectible scales. $110.

Watling SUPER HOROSCOPE, 1957. Watling didn't just knock the competition, they crunched them. They made a deal with Woolworth stores to place nickel SUPER HOROSCOPE scales across the country. Woolworth asked for handrails for the wobbly, and they got them. The big feature was a set of horoscope scrolls for each sign of the zodiac, with the scale holding 360 of them which were changed monthly. The dedicated kept going back every month for their latest horoscopes. $610.

Weight Check HONEST WEIGHER, 1987. It looks old, but it isn't. Weight Check, Inc. was formed in the late '80s in Arlington Heights, Illinois, a suburb of Chicago, to make computerized scales. 2000 SCALE offered weight and calorie analysis for 25¢, and diet recommendations for another quarter, total 50¢. They stuck the unit into a molded plastic reproduction of the old NATIONAL scale that gives a readout in pounds and ounces while music plays. In blue, red or yellow. Out of business. $160.

Building a Collection

Some years back, maybe as late as the early 1980s, when you asked a coin machine collector or dealer where they got a certain machine they were quite likely to say, "In a barn," "In an out-of-the-way antique shop," or, "Under the elevator shaft of an old slot machine factory." I didn't make this up, and machines have been found that way. I have made finds in the basements (rat infested, I might add) of old skid row buildings that had once been arcades, in run down factory warehouse sheds, in junk yards and on a Chicago commuter train, the latter leading to discussion that led to a discovery. Anyone who has been into vintage coin-ops for any length of time will have a pocketful of "finder" stories, and someone once suggested that a book on the subject of finds would be interesting reading.

for trade stimulators and slot machines, not to mention jukeboxes and arcade machines. So finds are still being made, and some of them are stunning. It's just that they are being made in different places, and you don't hear about them as often as you used to.

One reason for that is the antique mall. When malls first started we had no idea that they would change the marketing of antiques so drastically. In years back all you had to do was take your station wagon (vans were in their infancy) and tool around the countryside and find stuff. This was particularly apropos for coin machine collectors. I remember one collector telling me about the trip he took along the Ohio River from Little Egypt at the tip of southern Illinois to the eastern end of Ohio, finding

Most vintage coin machine collectors have a little bit of everything. Here slot machines, a jukebox and vending machines are displayed with an array of antique advertising in Don Reedy's collectible rec room in Frederick, Maryland.

But when you ask someone today, the answers are likely to be quite different: "In a mall," "...at the Chicagoland Show," or the most likely, "I got it from a specialized coin machine dealer in St. Louis, Missouri," or wherever. Times have changed. The quick and easy conclusion is that all of the good old machine finds have ended, and we have flushed out all we'll ever find in the woodwork. While there might be an element of truth to that statement, it really isn't so. Perhaps there aren't as many machines readily available in the "finder" market as there used to be and most of the "easy finds" have been made. Yet major discoveries are still occurring. The amazing fact is that the finest peanut vending machine ever found was sitting in an antique shop in upstate New York late in 1992 and went on to garner big dollars from a specialized collector. The same is true

stuff all along the way. He virtually built a collection of trade stimulators on that one three day trip. His excitement at making his finds was infectious; he explained how he had gone into a dark, old antique shop at the bottom end of Indiana, where he spotted a tall turn-of-the-century poker machine along the back wall. Four shops later he found a bartop dice game, and when the trip was over he had five or six machines he hadn't had before. These used to be called "buying trips." The machines weren't in old barns or garages or basements, but right out there for everybody to see in antique shops. The important thing was that you had to find the shop, and go in and see what they had. It made any tour exciting. The prices were good, too. It's not that the dealers didn't know what the machines were worth. It was that pricing was in its infancy, and no one knew what vintage coin-ops should sell for. So even priced at what the traffic would bear, some machines were high priced, while others were dirt cheap. You had to be on your toes selling, and keep your wits buying.

Contrast that to today's market. For one thing, there is a price guide for everything. Not that they reflect true values. Often they are way off the mark, as it is difficult for any one person or a pricing panel to know everything. Particularly in a field where the coming and going fads are so volatile and rapid. But if something is listed in a book at $500, no matter what the condition, you can bet that a dealer has it at $625 or thereabouts to cover the possible palaver that accompanies coin machine buying. What is needed is a return to rational and realistic pricing. Dealers that have machines sitting around for years gathering dust at inflated prices aren't doing themselves any good when they can turn them into productive pieces if they are priced where the market can accept them and actually buy something. Same for the owner of a machine that wants to sell it. Just because a price guide says $500 doesn't mean that you can get that in a sale. People that have made high bids at auctions based on book values, only to find that they are stuck with the machines when they try to sell them for anywhere near what they paid for them, can attest to that problem.

Back to the antique mall, and the greatest change in coin machine finding, and marketing. When that antiquer of some years back made that trip along the Ohio River the targets were shops, and what might be in them. Nowadays it is a lot easier to go from one town to another and see antiques because there are malls in just about every area (would you believe that Spearfish, South Dakota, has *two*!). But there is also a price to pay for that convenience. Somehow we are visiting shops much more and enjoying it less, to paraphrase an old cigarette advertising phrase. In the past it was up to us to find the coin-ops in dingy shops, and as travellers we had an advantage over the dealers that had to stay put, or couldn't be everywhere at once. But these days, when someone comes up with a great machine, and takes it into their mall booth, bingo! Before you know it there are a dozen dealers gawking at it, and zip! zip! zip!, based on the opinions of some and the price book knowledge of others, it moves through four or five hands with a bump in value each time before a collector ever sees the thing. Sure, finds have been made in malls, but they are few and far between as the dealers in the malls are there first and foremost. A nice piece hardly has a chance to bring the excitement and flush of finding at a reasonable price to someone that will keep it. Do I miss the Good Old Days of "hittin' the shops?" You bet I do. I enjoy malls, don't get me wrong. But they are the Wal Marts of antiques, and have a way of blowing out the little old ladies that sell things from grandma's house. Is it any less fun finding machines today? No, it isn't, because antiquing is fun no matter when and how you do it, and finds are still being made. Good ones, too. But you've really got to beat the bushes in the countryside, or have a lot of knowledge and know something about a machine that no one else in a mall knows. More often than not you are pitting your expertise against a couple of dozen people that work hard at their craft and are frequently as smart as you are. The amazing thing is that if you want to collect something—anything—you can start the collection today and be well in your way in a few months. Malls can help you do that faster. But...there was a time when you could pass a barn with a sign "Antiques for Sale" that wasn't there the week before, and would be gone in a month, and find something that became one of the treasures of your collection. These may be the good old days, but there were some then, too.

Let's assume you want to become a vintage coin machine collector. How should you go about it? You can buy the machines at antique coin machine shows, in antique shops or malls, or from a friend. Or you can reach out and find your own, and in the process maybe get a good buy or even make a historic discovery. The dyed-in-the-wool collectors prefer the latter route as it gives them some latitude over their buys, and adds a certain degree of adventure to the hunt. When someone asks a collector "Where'd you get that machine?", they'll be more likely to be impressed with your answer if you found it on your own than if it was ordered by mail from a dealer in vintage slots.

Not to knock the dealers or the shows. They perform an invaluable function by providing clean, shopped machines with a "home," with the sellers standing by their sale if something goes wrong or isn't as presented. For many, that safety alone is worth whatever they pay for a machine, and the price is usually right as dealers are in competition with each other so have to stay within some bounds. There are also price guides that keep them honest, and give buyers some assurance that they got a fair deal. There is little question that in the buying and selling of vintage coin machines there is a direct relationship between the price, or estimated value, of the machines and the transaction. But price is not as influential a determining factor as you might believe. The key factors in buying and selling are the reasons why, and at what point in the distribution stream you take action, the latter being the most significant influence on the money expended or received.

In the field of vintage coin machines, as with any fine antiques, there is a rigid producing and marketing stream which is paralleled by an equally rigid stream of supply and sales to collectors, just as there is in manufacturing and distribution. The parallels are uncanny. In coin-ops it is customary for the pickers and dealers handling the piece to each sell it for a profit. For ease of understanding, let's assume that the sale price is always twice as much as the cost. This traditional "Dealer Doubling" is deeply imbedded in vintage coin-ops sales and has become the way of its world. It means that a machine that was "discovered" for $100 will quickly run up to $1,600 in estimated value by the time it has reached its fifth owner, from original owner to a picker at an estate sale, to a flea market, to be found by a specialized coin-op dealer and sold to a collector who resells it to an advanced collector, with the cost going even higher if the machine is rare and unique enough to stay the course. In the case of a particularly interesting or exciting find, and because the dealers and advanced collectors are in such close contact with each other, that can happen as fast as a week or two or three through myriad telephone calls, actual visits to view the machine and subsequent wheeling and dealing, a buying and selling characteristic of the rarer and more desirable coin-ops.

There is a great value to staking out a position on the coin-op distribution stream and sticking to it in order to get what you want. Some buying examples will prove the point. Assuming you are a collector that has a good eye for rare and exotic machines; by staking your claim as a buyer at "Top Retail" you can save enormous amounts of money and travel time by just waiting for the "finds" to work their way up to you, at which point you make the buy with little question. If you are that kind of a buyer, your task is simple. Just make contact with the specialized dealers across the country, tell them what you are looking for, and the machines will come to you. You will also want to attend the advanced auctions of coin-ops and antique advertising. You might end up in a bidding war with other collectors, both with dealers and at auctions, but you still have a shot at a machine that would otherwise be lost to you. At that point you make the decision about how much you will be willing to pay for what you get, passing to other collectors if their buy decision leads to a willingness to make a stronger commitment.

The same goes for the advanced collectors. They often maintain contact with specialized dealers as well as collectors to pick up what they want. They have usually heard about any new machine on the market through the vintage coin-op underground conduit. If they are interested, they make their wishes known and can then pick and choose what they want, provided they are willing to pay the price for it.

All of this is at the higher levels of vintage coin-op collectibility. But what about the rest of us? If you want to get good machines at good prices, you should frequent those steps in the marketing stream that offer you the prices you are willing to pay. Unless you are willing to get into a van and scour the countryside, hit farm and out-of-the-way estate auctions, and dig through junkyards and the like (which some people like a lot, often leading to very good buys), forget being a picker and a finder. Your best best is at the next entry level at flea markets, antique shows, antique dealers in malls and shops and advertised auctions. In all candor, this is where most of the vintage coin machines that ultimately enter the collector market are found. But you must face the fact that they might be a bit rough, non-working and even "basket cases," needing some TLC and an experienced hand to get them back into shape and operation.

That brings us to the specialized vintage coin-op dealer, and the most reliable source for collectible machines. In the main they are dependable, trustworthy, informative and, most important of all, willing to stand behind their sale. They have to know their product to be able to buy wisely, "shop" the machine to bring it back to workability, and arrange for restoration where needed, or where desirable for a display machine.

You will find these dealers advertising in the leading antique publications, in local newspapers (when ads say "Will Buy" it usually means the advertiser will sell, too), in the coin-op fanzines (specialized coin machine collector magazines) and exhibiting at the many vintage coin-op, jukebox, slot machine and antique advertising shows that take place around the country throughout the year. To get the needed assurance of their reliability, confirm the fact that they will give you some form of operating guarantee with your machine and then ask for references. There should be no problem in asking for the names (and telephone numbers) of recent customers. And once contacted, these happy (or perhaps even unhappy) customers will fill you in in a hurry.

You should pick your vintage coin-op dealer in the same way you would pick your doctor. If you intend to become a coin-op collector this could well be a long term relationship. The place to meet the specialized coin machine dealers, and even other collectors, is at the world class Chicagoland Show conducted twice a year in the Spring and Fall, usually in April and November, or at the half a dozen or more shows around the country running throughout the year. Keeping track of vintage coin-ops through the fanzines, or watching your local newspapers (check the antiques shows section weekly!), will alert you to these events. You'll get a crash course in coin-ops and collectibility when you go, and will most likely enjoy every moment of the visit.

What about going it alone? Can you really collect coin machines without going broke? And can you protect yourself against dumb mistakes? As in all collectibles and fine antiques, knowledge is power. It is also protection, and an enjoyment enhancer. So the more you know the more you will enjoy your new found

Some enthusiasts specialize in groups of collectibles. Phill Emmert shows off his pinball games and arcade machines in his Colorado Springs, Colorado home. Left to right are Gottlieb RACK-A-BALL (December 1962), Williams LINE DRIVE (March 1972), Williams JUMPIN' JACKS (April 1963) and Bally BUS STOP (October 1964) pinballs and Chicago Coin ALL STAR BASEBALL (January 1963), Williams PITCH & BAT (January 1966), Williams OFFICIAL BASEBALL (February 1960), Williams BATTING CHAMP (April 1961) and Chicago Coin BASKETBALL CHAMP (March 1947) arcade machines.

enthusiasm. Buying through a dealer doesn't match the thrill of finding your own machines. But the moment you take that step into the unknown you have to face the fact that "you get 'em as you find 'em," and the machines you pick up are liable to be broken, missing some parts and might even need a lot of imagination to decipher their original appearance. You run just as much risk that they might be in great shape. It all depends. If you elect to go out and find your own, you should be well versed in the Three A's of the dedicated machine finders: advertising, auctions and activity.

Take advertising first. It is one of the classic ways that collectors and dealers find their machines. If you thumb through the pages of the collector magazines and newsletters you will often see "Wanted, Old Slot Machines" ads, some even saying "I'll Pay More. Call Me Last!" In the terminology of the advertising trade, ad groupings in that manner are called clutter. Every one of these advertisers gets results from their ads, or they wouldn't keep running them. Even if there are 7 or 8 ads looking for machines, most of them will get called when someone who gets the publication and reads the ads has a machine for sale. The problem from the buyer's point of view is that a competitive situation is set up from the very beginning, with the dealers bidding against each other, although in the final analysis someone does get the machine. Why not place your advertising where others don't.

If you live in a rural community, or one of the smaller cities, the chance of your newspaper ad being the only one looking for machines is very high. You are in command, and if there are machines in the community that people have (and they are found everywhere across this country) and might consider selling if they saw an ad, your advertisement might be the one to kick them over. Don't try and make a killing, or "steal" the machine. Just about everybody goes to antique shows and malls, and most people know how to find values in a price guide. Even public libraries have them these days. The fairer your offer the greater chance you have of getting the machine. In fact, most people tend to value their machines *higher* than their actual worth, so quoting (or even bringing) a price guide is a pretty good idea. But don't offer what the price guide says; go for about half. That's a fair and reasonable wholesale value, particularly if the machine needs some work. If you really want the machine and are willing to pay more, edge it up. But be aware that you are pricing yourself out of its true value. In the long run, "Advertising Pays!" Consistent advertisers consistently find machines, so it works.

The second "A" is auctions, the wild card of machine buying. You'd better know how to conduct yourself at an auction, and also be aware of what a machine is really worth before embarking on bidding up process. The best bet is to write down the amount you would be willing to pay for a machine, and stop your bidding at that point. But start low. If you are all alone in the crowd, and obviously the only coin machine collector, you are liable to get a bargain. If you are among vintage coin-op collectors be wary as the bidding will climb faster than you can imagine. Here's the wild card part. Let's say that someone just decides they want that machine, and they'll pay whatever it takes to get it. Be careful about that, because such a bidder can run you up farther than you want to go.

Taking these warnings to heart, you have a better chance at estate sale non-consignment auctions to get machines at low costs than just about anywhere else. Often, truly unique and historic machines show up on the auction block after being hidden away for years in an attic, basement, barn, garage or old store. If you know machines, rural and otherwise out-of-the-way auctions are your best bet for making significant finds at the right price. Equally, they are the easiest way to overpay once you are swept up in the emotion of a bidding war between two or three people who each want that machine as bad as the next guy. Just be careful.

Then there is the most exciting machine finding activity of all: activity itself. Talk to people. Drive around, look around, never give up. Check out the "Amusement" and "Vending" families in your area in old town and city directories, and track down the people. Pop a barn or two. Pass out business cards that say you are looking for old coin machines. Write letters like crazy to people who were in the business. You might even check out the local museums to see if they got any confiscated machines from the local police, and never displayed them. Every one of these activities has located machines. That museum shot is as far afield as you can imagine, but it has worked two or three times that I know of.

Legendary finds have been made following these proven techniques. And some funny stories have come out of them, too. After zero responses for over six months, in 1992 a Minnesota collector got an answer to his newspaper ad that led to the acquisition of a rare Mills floor slot machine from 1904. He didn't steal it, but the price was advantageous, and he added a marvelous machine to his collection. A Ohio collector went to an auction in a small farm town after getting a flyer that illustrated an 1890s dice machine that he had never seen before. No one had, in fact. It was a true rarity. He showed up at the auction as the only collector over an hour before it started, licking his chops. Five minutes before the gavel first hit the block, about 7 or 8 other coin-op collectors showed up, and an old trade stimulator that sat in the middle of an auction where nothing went for over two dollars suddenly shot up into the thousands, with the locals shaking their heads. Things happen for the better, too. A telephone call to a local librarian in a small Indiana town that once had a vending machine factory led to the name of the surviving family member, who was thrilled to be contacted and share her reminiscences and sell the three machines that were left. It can be done, and it's fun.

Collectors dedicated to a single area of expertise often cluster their machines. Trade stimulator and counter game collector Bill Whelan of Daly City, California puts them on glass shelves in display cases. Labels track the developmental histories of the machines, displayed by families.

Maybe the shoe is on the other horn, and you're in a selling mode and want to clean out the garage or basement or that stack of small vending machines that Grandpa left up in the barn loft. What should you ask for them? It's the same situation in reverse. Whoever the customer may be—dealer, coin machine collector or friend—you'll want to unload everything you've got, and you want to get top dollar. You're looking for the classic stock market position: sell high. The sad fact is that you can't be both things to the same machine. Assuming you don't know what these machines are really worth, and realizing you are in a vulnerable position, the thing you probably fear the most is that some slicker will come down the pike and drop a few bills on you, only to have you learn later that you gave away the store. If your buyer knows something you don't know, you could be making a costly mistake. That gets you right back to the classic stock market leveler: knowledge is power. The answer to this dilemma is to gain the knowledge you need to make the buy or sale you want. Knowledge goes two ways. What you need to know to buy a coin machine is exactly what you need to know to sell it. You may not have looked at price guides in the past in this manner, but that's why they're there. There is also one at the end of this book to help you in both the buy and sell situation. Not that price guides are the final word. The true collectors and serious dealers know better. Price guides are out of date the moment the ink is dry on the paper and even before the binding is finished, for the simple reason that the value of fine antiques fluctuates with the weather, calender, the value of the dollar, the price of bread and any other financial and emotional factors that may be in the wind. Antique prices are rarely static, and coin machine pricing is more volatile than most.

look for a discount store to do better. Same for antiques. If you notice, most busy antique shops tend to price their goods about 20% below the price guide values. It rewards them in a way that keeps them going; it brings in business. So if you're buying, use a price guide as a guide to what others think something is worth, and figure your cost will be the listed price or a little less based on the conditions indicated. But if you're selling, take that price guide cost and cut it at least in half, more likely into a third, and figure that's what you'll get if you are able to make the sale, because if you think you're going to get a lot more that machine will probably start gathering dust before it gathers any attention on the part of a buyer. If your buy and sell actions stay within these guidelines you're doing pretty well.

Conversely, if your insurance company wants to know what your machines are worth, realize that time is money and quote the price guides religiously because that is what it will probably cost you to replace something right away if you aren't willing to start all over and find your machines one at a time, with a story behind every acquisition.

How can you tell a good machine from a bad one? The fast answer is that there are no bad machines, although some machines are worth a lot more than others. The thing to look for is

Collector/dealer Bernie Gold of Great Neck, New York has created an aura area of his favorite machines, including an original green cabinet Caille Bros. CUPID which he has had for about twenty years. Artifacts include slots, vending and music machines amid a wealth of antique advertising.

But that's not the worst thing in the world, and if you view moderately current price guides as their name suggests—-guides—you'll be on fairly solid ground in determining value. Few things double in a year, and even fad pieces take a slow climb if there are enough of them around. So something valued in a price guide at $800 last year isn't going to be flying off an auctioneers table at twice that this year, or probably next. Unless it's a rarity and people didn't find out until just now. But in the main, the pricing is just about as stable as the currency with fluctuations every 3 or 4 years enough to keep up with the trends. One good piece of advice is to regard antiques like any other business, and consider a price guide value as list price. Right off the bat, if you were planning on buying a refrigerator, you'd

generally a combination of things. If it's vending, look for old, generally wood with a colorful decorated front and an inside clockwork mechanism. If it's an automatic payout slot machine or trade stimulator, find out if it's cast iron, the cabinet construction of choice between the turn of the century and 1922 (to be replaced by aluminum). Most good finders carry small refrigerator magnets in their pockets so that they can check the cabinet of a machine on the sly and determine if it is iron or aluminum. The magnets stick to iron. They won't stick to aluminum or pot metal. And iron is the magic metal in payouts, the more elaborate the better. In pinballs and baseball games and bowlers it's the backglass. If the glass is worn or scratched or broken forget the game, as over half the value is in the "cosmetics." In juke boxes look for bubble tubes or decorated glass, items which place them right in the middle of the Golden Age circa 1940-1955. In scales look for elaborate cast iron. If your finds or barn storage match these specs it doesn't mean they are barnburners, but their chances at being a better than average coin-op are greatly multiplied. Happy hunting.

Dedicated gumball and peanut vending machine collector Stan Harris of Philadelphia, Pennsylvania displays his machines on long shelves that show the progression of development. It is often mind bending to see such rare objects so comfortably and engagingly displayed.

Auctions: When collector Stan Muraski of Rockford, Illinois saw this pile of broken iron held together by string at an auction he had to have it. Price was relatively low, less than $400. Name on the side of the casting identifies it as the EMPIRE drop card machine made by the Automatic Machine Company in New York City in 1888, making it the oldest cast iron card machine ever found. Next is restoration. $4,515.

Advertise: Trade stimulator collector Tom Gustwiller of Ottawa, Ohio, advertises widely in collector and coin-op magazines. It leads to finds. Some are puzzlers, such as this old nickel play wooden cigar wheel in marvelous condition. An obvious take off of the Decatur FAIREST WHEEL, its maker and date remain unidentified. $2,810.

Advertise: Collector/dealer Frank Zygmunt of Westmont, Illinois is a frequent advertiser. A summer 1994 caller alerted him to an old pool hall built in 1905 that had some coin machines in storage. Among the finds when he followed up the lead was the first known example of the Chicago made O. I. C. VENDER peanut machine previously only known from advertisements of around 1916. The cash box is in the top. A major find! $12,520.

Activity: Typical of flea markets and country fairground antique shows all over the country, this Exhibit Supply VITALIZER of 1939 hung around half a day before somebody picked it up for $150, chiseled down from $200. Fixing up and replacement paper will make it an interesting and more valuable arcade machine collectible. And it does relax tired feet. $260.

Auctions: When a trade stimulator collector spotted this dicer in a rural farm auction flyer in 1990 he knew he had to have it. The picture was bad, and the auction was far out in the country. Yet a number of coin-op enthusiasts showed up. In an auction where everything went for 25¢ to a buck or two it ran up well over $10,000, and left the locals in awe. Marked a tantalizing E. A. M., and that's all, it remains enigmatic and the only one of its kind known.

Activity: It took a lot of vision to go for this one. It was in the parking lot at the Chicagoland Show in April 1994, missing parts and glass. A young collecting couple got it for $50. It turned out to be the only know example of the all but unknown Chicago Coin SCALE-O-MATIC combination horoscope scale and vibrator of the late 1940s. Next comes major restoration. $460.

Activity: There are coin machines everywhere. The question is, are they collectible? 50 years from today this air machine will be an interesting piece. 100 years from today, if any survive, it'll be a treasure telling a lot about public transportation in the late 20th century, when people had their own cars and trucks. $90.

Buy carefully: It looks like a slot machine, but it's not. It's a toy. These fake slot machines were sold in airports in the 1970s and '80s, particularly in Nevada, and are nothing more than dime banks. Numerous budding collectors have put good money out for these non-collectibles...unless you're into modern banks. $50.

Activity: Just because you make the biggest coin machine find of your life doesn't mean it is a collectible worth restoration. This triple coin chute 75¢ ice machine, the size of a railroad box car, sits on a roadside in Michigan. Unless something as inconceivable as a dedicated coin-op ice box collector comes along, it'll stay there. No value.

203

Buy carefully: While this so-called "Mills" GOLDEN NUGGET machine looks great, and plays well, it is a fantasy piece. They were never run by the Golden Nugget Casino. In fact, the Golden Nugget has asked the makers of these machines to cease and desist. They were created by coin-op collectible dealers as revamps of old Mills slot machines, with many of the later versions brand new from the ground up. They're fun, but they aren't antiques. $1,610.

Buy carefully: This attractive "Jacks" machine looks like a period piece, but it is new. They were first made in the 1980s from the original 1930s patterns, and have been widely sold. And why not, they play well, and are less expensive than an original. A clue to their identification; they play on a dime. The originals were penny or nickel. $360.

Buy from a dealer: Elaborate details, such as the original nameplate on the EXCELSIOR race game made by the Excelsior Race Track Company in Chicago in 1890, are most often missing from newly discovered machines. Prime pieces, such as this, often end up in dealer hands. If you want machines in that shape, shop for them.

Buy from a dealer: One thing you'll find at a dealership is selection, with machines sold as represented and backed by some form of assurance against damage. They also come from a "home" for later maintenance and repairs. This is the jukebox and speaker display room at Oldies But Goodies in Oroville, California, run by John and Thil Wilcox.

Buy from a dealer: It usually takes a dealer willing to risk capital, and some restoration costs, to pick up incredible pieces such as this brightly decaled Mills LITTLE PERFECTION "Round Top" card machine of the early 1900s. If you want a shot at items like this, get close to a dealer with access to them. And is that "What, me worry" Alfred E. Newman of MAD magazine at the right?

Buy from a dealer: Many dealers invest a great deal in inventory, and show off their finest at shows. Larry Lubliner of Re-finders (far left) displaying his line of saloon artifacts, trade stimulators and slot machines at a Las Vegas trade show. He is talking with a potential customer.

Resources

You can learn a lot about vintage coin machines at some of the national and major regional shows. In booth after booth you will find the very things you most enjoy. A tour of the floor (for the really big shows, like the twice-a-year Chicagoland Show, it usually takes two days to really hit all the booths) will in all likelihood make you an experienced enthusiast, even if you never buy a thing. The show itself is worth the price of admission, whatever the price. You will see what's hot, and what isn't, and get a very clear idea of the pricing of the various forms of machines. You will also learn what it means to restore, and over-restore, by the testimony of your own eyes. And best of all, the exhibitors are just as keyed up about their exhibits as you might be, and are always willing to talk at length about the hobby, their machines and your collection. It is a canned education, and worth any trip you have to make to participate as a looker or buyer, or both.

Shows provide an opportunity to go shopping for machines, giving you a chance to see them and negotiate a purchase. A wide selection of jukeboxes, counter games, venders and all other forms of coin-ops are on the floor. Getting there early gives you the best chance of getting what you want as the machines move out the door quite fast.

The Chicagoland Show (the largest in the world serving the hobbies of vintage coin-ops, antique advertising and saloon artifacts) is held twice a year in the spring and fall. Local shows are also held around the country throughout the year. Subscriptions to coin machine collector magazines will give you all the show dates.

There are a lot of vintage coin-op shows around the country that will have jukeboxes, slots, counter games, pinball games, arcade pieces, scales and vending machines on display, with everything for sale. If there is a local or national show in your area, and you have even the slightest interest in these machines, by all means go. See for yourself what exciting collectibles vintage coin-ops can be. In the case of the spring and fall Chicagoland Shows, make that world class. There are far more Europeans and Pacific Rim and North, South and Central American visitors to this world class show than there are Americans at any of their shows (which are very few and far between), which leads to a polyglot crowd. The show runs twice a year, usually in April and November, in at the Pheasant Run Resort in St. Charles, Illinois, a western suburb of Chicago. For information about show dates, call Steve Gronowski at 708-381-1234.

The colorful front of the Pace DANDY VENDER of 1931 calls out for a better award card than the handmade one now in place. It is likely that the machine ran this way in its later years after the original card wore out. This award panel has been reproduced and can instantly upgrade the machine. That hand painted period job is interesting, however, and should be kept, maybe framed. $360.

Keeping a slot machine in running order isn't as tough as it looks. The coin moves through the machine like water down a river, bending and turning where required. It is all done very smoothly, with overlapping jams and missing springs usually the biggest problem. Manuals are available for most slot machines, and you can probably keep things working by yourself. If all else fails there are restorers and dealers that handle maintenance and repair. This is a Jennings SUN CHIEF TIC-TAC-TOE, typical of pre-1960s mechanicals. $1,610.

This is an Automatic Machine Company AUTOMATIC DICE, made in New York City in 1892, as-found. While the purist doesn't want to lose the original paper award panel, a matching reproduction without the scratches would add to the appeal of the machine. The trick is to somehow retain the original (framed under glass would be nice) while upgrading the machine with a careful restoration.

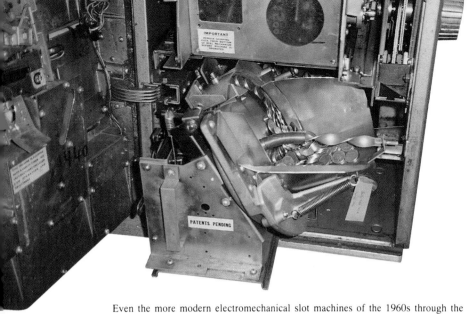

Even the more modern electromechanical slot machines of the 1960s through the '80s can be maintained at home, although they are a little more complicated. Manuals are available for all Bally machines, such as this early MODEL 742A MONEY HONEY, and its later variants, as well as for Mills, Jennings, Ainsworth ARISTOCRAT and other better known machines. Other lesser known electromechanical and solid state slots are orphans and have no maintenance backup material. You are safe sticking with Bally, Mills and Jennings. $1,810.

The ultimate in restorations. When Burton A. Burton (right: of Malibu, California, Founder of The Casablanca Fan Company) discovered his Watling DERBY in the early 1970s, all he had were some of the castings from the top with none below his hand level, plus a catalog picture of the 1911 original. He restored the only known example from the floor up, making a new mechanism, castings and the cabinet. When Larry De Baugh (left: president of Novatec Packaging in Baltimore, Maryland) found his DERBY after a junkyard fire in California in 1989, all the wood was missing, although he had the mechanism and castings. He restored his from the same catalog picture. When the two DERBY machines were compared upon completion they were only 1/8" apart in dimension. Value of the two exceeds a quarter million. Their owners say each one is inestimable, adding they are not for sale.

Two "eastern shows" also pull the crowds in the spring and fall. The first is the Philadelphia Game Room Show, held at the N. E. National Guard Armory in May, with the second billed as "The largest East Coast show of its kind," at Hackensack, New Jersey in October. For dates and data call Bill Enes at 913-441-1492. And last, but certainly not least, is the California equivalent, the Coin-Op Super Show. It is held at the Pasadena Convention Center in the spring and fall, and periodically in Modesto, in northern California, in the fall. For show dates and information call Rosanna Harris at 303-431-9266. Hit the floors of these shows and you will be a hooked coin machine collector in no time.

The next thing to do after that if you are still interested is to subscribe to one or more of the vintage coin machine collector magazines, both to learn about the machines and the many sources of supply available to you for replacement parts and maintenance items and, when needed, complete restoration capabilities. All of this information is at your fingertips through the fanzines of the vintage coin machine hobby. It is an enthusiasm blessed. And the reason is because it is also an enthusiasm splintered. Coin-op collections are often polyglot. But the media isn't. Because the various groups of machines are so different in appearance and application, each has garnered a hard core following, and the fanzines reflect these vertical disciplines. Many, if not most, coin machine collectors get them all, or at least 2 or 3, to keep up with the hobby (or hobbies, if you will) and not miss a thing. For general horizontal coverage of all forms of vintage coin machines, and a high level of editorial interest, there are two prime fanzines:

Coin-Op Classics.
17844 Toiyabe Street, Fountain Valley, CA 92708
Bi-monthly. $48 per year US ($US54 Canada).

Covers all areas of coin machines, from slots to vending, jukeboxes to pinball, arcade to counter games, and even scales. This is the most editorially filled fanzine in the hobby, profusely illustrated, including 4-color. Covers collecting, coin machine history, recent "finds" and much, much more. The leading generalist magazine.

Gameroom.
1014 Mt. Tabor Road, New Albany, IN 47150
Monthly. $24 per year US ($US30 Canada).

This has long been the bargain of the hobby, 12 months for 24 bucks. Also a generalist publication, with concentration on jukebox, arcade, soda machines and pinball, not to mention antique advertising, Coca-Cola and gameroom stories. Format is B&W on offset stock, with 4-color covers. Plus lots and lots of classified ads.

For vertical coverage of specific areas of coin machine interest, these publications are fascinating:

Jukebox Collector.
2545 SE 60th Court, Des Moines, IA 50317-5099
Monthly. $30 per year US ($US36 Canada).

If you like jukeboxes you will love this magazine. Every issue has historical jukebox stories, plus restoration tips, news of shows, and a great looking jukebox on the cover in color every month. Again, lots and lots of classified ads.

pinGame Journal.
31937 Olde Franklin Drive, Farmington Hills, MI 48334.
Monthly. $28 per year US ($US32 Canada).

B&W. Pinball only. Mostly current and newly introduced games, and some historic. This is a true insider publication, and if you like pins you will be in on the know months before you see the games on location. And wads of classified ads selling and buying pinball games, plus ads for parts, manuals, everything.

Coin Drop International.
5815 W. 52nd Avenue, Denver, CO 80212
Bi-monthly. $12 per year US ($US18 Canada and offshore).

Here's the new bargain of the hobby, a full year of 6 issues for 12 bucks. Large tabloid newspaper format, and the latest publication to join the fold. The first issue came out in July 1994, with issues coming out every-other-month after that. Major concentration is on vintage slot machines, and the states allowing collectibility, although all phases of coin-ops are covered.

Many as-found machines cry out for restoration. This is the original condition of the Krema MATCH VENDER of 1922, illustrated in the vending machines chapter, before its restoration. Rust and graphics damage require restoration in addition to the working mechanisms, with the cosmetics almost more important to collectors.

If any of these fanzines seem to cover your area of interest I highly recommend that you get a subscription and quickly get yourself knee deep in your area of collectibility. You'll find it rewarding. Or get them all. You'll learn a lot in a hurry.

Once you have a machine or two you no doubt will accept the dictum of the hobby: they have to work. Most antiques don't have much expected of them. More often than not they just have to sit there and look pretty. Not so with mechanical antiques, and vintage coin-ops in particular. They not only have to look good; they also have to do what they were made to do. If it's a slot, it has to pay out when a winner is hit. Soda machines have to vend bottles, cold. And pinball games have to give you the kinetic playing experience of the original models just as they did when they were popular. Coin machines face some of the toughest specs of any antiques. To be regarded as a valid collectible they have to work. Otherwise they are just dust collecting eunuchs unable to perform and do not have much value.

That having been said, what do you do when you find a machine that has parts missing or doesn't work? Simple. You bring it back to working condition. But working alone isn't enough. The machine also has to look like it did when it was operational, which means that its cosmetics need to be retained, restored or replaced. This need to fix up, replace missing parts, restore where needed and recreate the experience of the original play or operation of coin-ops is at the very heart of their collectibility and value. That is why there is such a great fluctuation in values between a "basket case," a so-so find and a mint working original. Short of looking good and working, someone along the line has to bring the less than pristine machines back to acceptable standards. And that takes time and money. To meet this need, a whole new cottage industry has grown up around vintage coin-ops to supply the missing parts, provide reproduced material and restore the machines back to original condition. Sometimes the latter is taken too far, with over-restorations making these venerable commercial machines into something they never were when they were money grabbers. That in itself is an excess and does as much to detract from the value of a machine as the frequent as-found rough condition of an original.

If you have a vintage coin-op and face the need to improve its appearance (outside restoration) and workability (inside restoration), some basic ground rules apply.

Outside Restoration

1. Dust the machine. The best bet is to start with vacuuming. Use a strong vacuum cleaner with a detachable brush head to do the overall cleaning and corner work. You will be amazed at how much dirt and grime will come off at this simple stage. If there are any award or instruction cards or other papers or decals on the outside, remove or protect them if possible so that they don't crumble and get sucked into the vacuum cleaner, as that would be a loss of major proportions.

2. Clean the machine, but be very careful when you do so. Modern detergents, industrial and household cleaners are so strong they can eat away or totally destroy the very cosmetic features you are attempting to retain or restore. Plain lukewarm water and a soft brush, maybe with old original Ivory or Castile soaps to get at the real grime, should do the job. The stronger petroleum based cleaners of today can easily remove paint and printing from most cabinets or paper.

3. Evaluate and repair the finish. If there are chips, consider minor retouching, but match the paint perfectly on a sample before you do so. Modern paints don't chip (they can peel, but they won't chip) and rarely match the strong solid colors of the past. Whatever you do, do not repaint a machine unless it is a totally hopeless rust bucket. Advanced collectors far prefer the original finish, chips and all, over a stripped, polished or repainted cabinet, and they can spot the differences right away. Finally, original paint has more value than a repaint. But if you do have to repaint, find another original, or track down an old ad or spec sheet, so that you get the colorations correct. Anything less than exactitude is a fantasy piece.

4. Retain, restore or replace the paper and other cosmetics. Collectors go ballistic with joy when a machine find is made with the original paper and decorated glass intact, because these artifacts are far more vulnerable to the ravages of time than the mechanism. Such adoration of minutia is typical of coin-op collecting, and a key determinant of value. When the paper and glass details are torn, rotted, broken or missing be aware that you have a machine of greatly reduced value. Many such items have been reproduced to meet the need of exacting restoration. While not cheap, they can go a long way toward bringing a machine back from oblivion.

5. Provide proper access to the machine. If the original lock and key are on the machine you are way ahead of the game. A replacement is almost as good, but not quite. If you have a locked machine and no key, try a bunch of old keys. One just might fit. If not, you may have to drill out the lock and put in a new lock and key. Always keep a key in the machine, locked or not, and a dupe somewhere else.

Inside Restoration

1. Try and find a manual. You are about to get involved in the nuts-and-bolts of mechanical systems, and quite often, electricity. So a working maintenance and trouble shooting manual, and schematics where electric circuits are involved, will be a great help. Even if you don't know how to use them yet, you will probably learn by doing. Many coin machine manuals have been reproduced in books, while originals can often be found or borrowed from other collectors. Manual, or no, remove the back to get inside. Then, before you remove any parts or components, take pictures or a video, and repeat the picture taking at every significant disassembly step. You will be glad you did. Disassembly and reassembly may look easy, but having a few pictures or a video showing how your machine came apart to guide the way to put it back together again never hurt.

2. Clear the coin path. Get rid of old lint and paper jams, and clean the interior with gasoline or lighter fluid—-don't smoke!—-and a tooth brush. Four out of five times, when a machine isn't working, the problem is in the coin process, be it a coin chute "gooseneck," coin escalator, coin slide or coin acceptor. Jams or coin lapping are a major problem. The judicious loosening of some screws, and the unjamming of coins, will often solve the problem and get the machine working again. Use toothpicks to poke at things because you do not want to damage the coin chutes and systems with a screwdriver or other piece of metal as small nicks and scratches can create even more problems for the future. And take a good look at any old coins that might have jammed a machine years ago. Some great numismatic finds and rare dates have been discovered that way. If you come across any bent coins, spend them fast and keep them out of your machines. They cause more jams than anything else.

3. Clean out the machine. In slots, many vending machines, scales and trade stimulators, the mechanisms often slide out once retaining hooks have been released. In machines with electrical systems, such as jukeboxes, arcade games and pinball machines, some components can come out, while most stay in, attached to the cabinet. Either way, clean out as much as you can, and give the insides a good vacuuming. But this time put a small cloth filter behind the nozzle to catch any loose parts or small screws that you will need to put the machine back together. Many a key piece has been gulped by a vacuum cleaner without such protection. Vending machines often come in for their own form of cleanout, as some of the products vended may still be there, gumming up the works. Old salted peanuts and other food products are particularly corrosive to aluminum and pot metal mechanisms. These must be removed with care to avoid stripping out part of the working mechanisms. Old, rusty mechanisms can also use a dipping in gasoline to break through aged grease and gum. Or you can clean components and gears with gasoline and that tooth brush again. But stay away from any solvent or cleaner when electrical connections are involved. WD-40, the magic ingredient in any mechanical restoration, can also be quite nasty, particularly when it comes to old rubber, fibre or leather gaskets and cushioning devices, because it eats them up. Paper too.

4. Oil things up. Any wear points metal to metal, pinions, bearings and the like need a drop or two of "3-In-1" oil. But keep it light as lubricating oils will stiffen up and clog everything. Heavy wearing parts need an application of Vaseline petroleum jelly (get it at the drug store) wherever the metal is worn and shiny to lubricate the parts. But if your machine has payout slides, *do not use oil or grease on them*. They require no lubrication whatever, and anything added makes them stick. Cleaning with gasoline once in a while is a good thing, though.

5. Try your machine. See if it works. If it does, stop fooling around with it. If it doesn't, start making some serious double checks. Are there any hanging, rusted or missing springs (you can spot the latter by unused lugs)? This problem is the most common for mechanical machines and fortunately just about every spring ever needed by a coin machine has been reproduced. You can get them from Bernie Berten, with his address in the following lists. Two stamps will get you his catalog with parts listed by machines. Are some electrical connections loose, or not made? If they look like they should be, or the schematic says so, reconnect. And take a look at any fuses, too. Many a machine has been stopped cold by a simple blown fuse, and that's all. Finally, are any parts missing? This might be a little tougher to determine, but you can often do it. If you have a manual, all parts are listed and numbered. Just check to see what's there, and what isn't. But in the absence of a manual, try and visualize the mechanical stream of the machine, and then envision the part when a connection seems to be missing. Quite often you can do it. Then what you have to do is find the right supplier, as many if not most of the parts that either break or are often missing, have been reproduced.

If you end up getting deeply involved with vintage coin machines, you will need to contact some of the major suppliers and experts somewhere along the line. Here are some people that can help you.

Vintage coin machine identification and evaluation

Richard M. Bueschel
414 N. Prospect Manor Avenue
Mt. Prospect, IL 60056-2046
Tel. 1-708-253-0791

Major shows and events

Chicagoland Antique Advertising, Slot-Machine & Jukebox Show
Spring and Fall
Pheasant Run Mega Center and Pavilion Building
Tel. 1-800-999-3319 (for rooms)
Route 64, North Avenue (2-1/2 miles west of Route 59), St. Charles, IL
Contact: Steve Gronowski (for show information)
Tel. 1-708-381-1234

Philadelphia Gameroom Show
Spring
Valley Forge Convention Center
Tel. 1-610-337-4000
1200 First Avenue, King of Prussia, PA
Contact: Bill Enes (for show information)
Tel. 1-913-441-1492
FAX: 1-913-441-1502

Collectors' Expo
Fall
Rotham Center
Fairleigh Dickinson University,. Hackensack, NJ
Contact: Bill Enes (for show information)
Tel. 1-913-441-1492
FAX: 1-913-441-1502

Coin-Op Super Show
Spring and Fall
Pasadena Exhibit Center
Green Street between Marengo and Euclid, Pasadena, CA
Contact: Rosanna Harris, Royal Bell, Ltd. (for show information)
Tel. 1-303-431-9266
FAX: 1-303-431-6978

Pinball EXPO
Fall (Chicago area)
Contact: Rob Berk
2671 Youngstown Road, S.E.
Warren, OH 44484
Tel. 1-216-369-1192
Toll Free: 1-800-323-FLIP

Book and Coin-Op Literature Dealers

Ken Durham Books
909 26 Street NW, Washington, DC 20037
Tel. 1-202-338-1342

Rosanna Harris
Royal Bell Bookshelf
5815 W. 52nd Avenue
Denver, CO 80212
Tel. 1-303-431-9266
FAX: 1-303-431-6978

Slot Machine, Trade Stimulator and Counter Game Parts and Restorations

Mike Andrews
Las Vegas Antique Slot Machine Company
4820 West Montara Circle
Las Vegas, NV 89121
Tel. 1-702-456-8801

Bernie Berten
Springs and Parts
9420 S. Trumbull Avenue
Evergreen Park, IL 60642
Tel. 1-708-499-0688
FAX: 1-708-499-5979

Craig Bierman
Speed & Sport Chrome Plating
404 Broadway
Houston, TX 77012
Tel. 1-713-921-0235

Fred De Baugh
222 E. Thomas Avenue
Baltimore, MD 21225
Day Tel. 1-410-789-4811
Evening Tel. 1-410-592-2936
FAX: 1-410-789-4638

Jack Freund
Slots Of Fun
P. O. Box 4
Springfield, WI 53176
Tel. 1-414-642-3655

Tom Kolbrener
St. Louis Slot Machine Company
2111-R, South Brentwood
St. Louis, MO 63144
Tel. 1-314-961-4612

Tom Krahl
Antique Slot Machine Part Company
140 N. Western Avenue
Carpentersville, IL 60110
Tel. 1-708-428-8476
FAX: 1-708-428-4471

Alan Sax
3239 Victorian RFD
Long Grove, IL 60047
Tel. 1-708-438-5900

Reel Strips, Decals and Award Cards

Evans & Frink
2977 Eager
Howell, MI 48843

Barbara Larks
The Gumball King's Decals & More
8444 Lawndale Avenue
Skokie, IL 60076
Tel. 1-708-679-4765

Bill Whelan
Slot Dynasty Coin Machine Restorations
23 Palmdale Avenue
Daly City, CA 94015
Tel. 1-415-756-1189

Jukebox Parts and Restorations

Bob Maller
The Classic Jukebox Works
1575 Cambridge Circle
Medford, OR 97504
Tel. 1-503-773-9534

Chuck Martin
CSSK Amusements
Box 6214
York, PA 17406
Tel. 1-717-757-3740
Toll free; 1-800-PIN-BALL

Funtiques
P. O. Box 825
Tucker, GA 30085-0825
Tel. 1-404-564-1775

Bob Nelson
Gameroom Warehouse
826 W. Douglas
Wichita, KS 67203
Tel. 1-316-263-1848

The Jukebox City
1950 1st Avenue, South
Seattle, WA 98134
Tel. 1-206-625-1950

Martin Richter
Jukeboxes Etcetera
223A N. Glendora Avenue
Glendora, CA 91740
Tel. 1-818-914-9434

Ben Humphries
The Jukebox Garage
1326 N. 22nd Avenue
Phoenix, AZ 85009
Tel. 1-602-253-8545

Jukebox Junction
P. O. Box 1081
Des Moines, IA 50311
Tel. 1-515-981-4019
FAX: 1-515-981-4657

Lloyd's Jukeboxes
22900 Shaw Road No. 106
Sterling, VA 20166
Tel. 1-703-834-6699
FAX: 1-703-834-9083

Steve Loots
Victory Glass Company
3260 Ute Avenue
Waukee, IA 50263
Tel. 1-515-987-5765
FAX: 1-515-987-5762

Pinball Game and Arcade Machine Parts and Restorations

Jim Tolbert
For Amusement Only
1010 Grayson Street
Berkeley, CA 94710
Tel. 1-510-548-2300
FAX: 1-510-548-8904

Chuck Martin
CSSK Amusements
Box 6214
York, PA 17406
Tel. 1-717-757-3740
Toll free; 1-800- PIN-BALL

Donal Murphy
Electrical Windings Inc.
2015 N. Kolmar Avenue
Chicago, IL 60639
Tel. 1-312-235-3360 (8:30 AM - 5 PM)
FAX: 1-312-235-3370

Bob Nelson
Gameroom Warehouse
826 W. Douglas
Wichita, KS 67203
Tel. 1-316-263-1848

Paul Biechler
Home Arcade Corp.
1108 Front Street
Lisle, IL 60532
Tel. 1-708-964-2555
FAX: 1-708-964-9367

Joel Cook
The Pinball Lizard
3270 East Fifth street
Tucson, AZ 85717
Tel. 1-602-323-7496

Tony Miklos
Pinball Paramedic
1372 Taggert Road
East Greenville, PA 18041
Tel. 1-215-541-4167

Steve Young
Pinball Resource
37 Velie Road
LaGrangeville, NY 12540
Tel. 1-914-223-5613
FAX: 1-914-223-7365

Vending Machine Parts and Restorations

Ary and Rose Kothera
Antique Vendor Supply
P. O. Box 411
Middlefield, OH 44062
Tel. 1-216-632-0423

Craig Bierman
Speed & Sport Chrome Plating
404 Broadway
Houston, TX 77012
Tel. 1-713-921-0235

Victor Vending
900 Parker Avenue
Dassel, MN 55325
Toll Free: 1-800-762-8028

Soda Machine Parts and Restorations

Paul Biechler
Home Arcade Corp.
1108 Front Street
Lisle, IL 60532
Tel. 1-708-964-2555
FAX: 1-708-964-9367

Dan Cooney
Cooney Island Antiques
3083 N. California Street
Burbank, CA 91504
Tel. 1-818-569-0216
FAX: 1-818-569-0228

Bob Fowler
7522 N. La Canada
Tucson, AZ 85704
Tel. 1-602-575-8805

Bill Mock
2640 SW 29th Way
Ft. Lauderdale, FL 33312
Tel. 1-305-584-8958

Jeff Walters
Memory Lane Sodaware
P. O. Box 6239
Laguna Niguel, CA 92607
Tel. 1-305-584-8958

Soda Fountains

Terry Schy
American Soda Fountain, Inc.
455 N. Oakley Blvd.
Chicago, IL 60612
Tel. 1-312-733-5000
FAX: 1-312-733-3621

Coin-Op Scales Parts and Restorations

Bill & Jan Berning
P. O. Box 41414
Chicago, IL 60641-0414
Tel. 1-708-587-1839

Some quick tips about buying and selling vintage coin machines:

1. Just because a slot machine says "Copyright 1910" on its reels doesn't mean it's that old. The Mills Novelty Company and others put this marking on their reels well into the 1940s to indicate copyright protection of the symbols.

2. Cast iron machines are highly desirable. Bring a kitchen magnet along with you when you're buying to test the metal, or show a buyer if you're selling. If it sticks, it's cast iron, and possibly made before 1922. If it doesn't, it's brass, aluminum or pot metal, all of which were used later.

3. To make a machine highly desirable, it takes 3 examples, and a competition between leading collectors to have an example. A single example of a machine is not as desirable. Rarity rarely translates into value.

4. Just because a machine is expensive doesn't mean it's worth the price. GOLDEN NUGGET 3-reel slot machines have been selling for anywhere from $1,000 to $6,500 at any given point in time, yet most (if not all of them) are fantasy repros and are not ranked as true slot collectibles. So watch out! Read a price guide and learn the machines.

5. If a pinball game doesn't work, and it's a highly desirable game and the playfield and backglass look good, get it. The same goes for selling it. It can be fixed. But be ready to face "shopping" (getting the machine back in playing condition) costs anywhere from $125 to $350.

6. If a pinball or arcade game works but doesn't have a backglass, or it's broken, forget it. The cosmetics are 60% to 80% of the value of the game.

7. If any coin machine, such as scales, pinballs, juke boxes, arcade machines, trade stimulators or counter games, has broken or missing glass graphics, pass on it. Or, if you are selling, get what you can for the game. The cost of recreating the art will be far in excess of the value of the machine. The same goes for the paper or advertising panels on most vending machines.

8. Most pinball and arcade pieces are hardly worth the cost of moving them. For every hundred miles away from you their value drops 20% because once you pick one up you have to get it home safe and sound. Your best bet is to find stuff in the same town you live in, or close by. The desirable pieces have value; most of the others do not, or not much anyway. If you just want one to play, great. But be warned that the collectible pieces are well known and identified. It is hardly worth your time to pick up a non-working less than desirable piece a 4 hour drive away. That's an 8 hour round trip, so assume the cost of the machine is what you pay plus what you could have earned or accomplished in those 8 hours. The same goes if you plan to sell a pinball. The further someone travels to get the game, the less you will get, if anything.

9. Know your jukebox models. Some 45s from the 1950s, that look a lot alike, can range from $750 to $7,500 in value. And the "many miles away" rule applies here, too. Read price guides and be sure of what you are getting before you commit to a trip or a price.

10. Unless you are a mechanical wizard, and enjoy the work, pass on the "basket cases." They are often far more work than they are worth, particularly if they are fairly common machines.

Index

A.B.C. Coin Machine Company
  JOCKEY CLUB, 156
A.B.T. Manufacturing Company, 7, 8
  ELECTRIC SKILL GUN, 130
  SCRAM, 93
  TRIP-L-JAX, 33
A.C. Novelty Company
  MULTI-BELL, 48
  MULTI-BELL VENDER, 48
Ace Manufacturing Company
  SPECIAL AWARD, 67
Ad-Lee Company, 176
  E-Z BASEBALL, 178, 179
  TRY-IT, 152
  TRY-IT BALL GUM VENDER, 152
Advance Machine Company, 7, 8
  CLIMAX, 171
  CLIMAX CHICAGO, 171
  GLOBE, 174
  HILO GUM CLIMAX, 171
Advertising Scale Company (Hamilton Scale)
  MR. PEANUT, front cover, 2
Aerion Manufacturing Corporation
  CORONET MODEL 400, 80
  FIESTA DELUXE, 82
Ainsworth Consolidated Industries (Australia)
  ARISTOCRAT, 16
Air vender, 203
Ambassador, Inc.
  AMBASSADOR, THE, 83
American Auto-Machine Company
  AUTO-MACHINE, 114
American Automatic Machine Company
  AUTOMATIC DICE SHAKING MACHINE, 139
American Biograph Company, 123
American Mutoscope Company, 123
  MUTOSCOPE MODEL A, 11
  MUTOSCOPE MODEL B, 11
American Mutoscope & Bio-graph Company, 111, 123
  MUTOSCOPE, 110, 111
American Novelty Company
  SCULPTOSCOPE, 111
  WHITING VIEWER, 111
American Scale Manufacturing Company, 184, 186
  MODEL 300 FORTUNE, 193
  MODEL 400, 195
  MODEL 403, 195
  MODEL 530 COMBINATION SCALE, 186, 196
  MODEL 531, 186
American Talking Scale Company, 187
American Tobacco Company, 167
Amusement Machine Company
  3 JACKPOT, 13
Arkansas Novelty Company
  3-A-LIKE, 163
A.S.L. Sales Company
  ACCURATE SCALE, 193
Atari, Inc., 135
  ATARIANS, 108
  PONG, 108, 111, 134, 135
  SUPERMAN, 108
Atlas Games
  TILT TEST, 164
Atlas Indicator Works
  BASE BALL, 125
Austin, Nichols & Company
  FOXY GRANDPA, 170
Auto-Bell Novelty Company
  TILT TEST, 164
Auto-Photo Company, The
  AUTO-PHOTO NO.7, 130
Automat Games
  SILVER KING, 180
Automatic Battery Company, 6
Automatic Beam Scale Company
  AUTO SCALE, 188
Automatic Clerk & Advertising Company
  AUTOMATIC CLERK, 168, 169
Automatic Games
  SILVER KING, 180
Automatic Games Company
  FIFTY GRAND, 95
Automatic Industries, Inc.
  WHIFFLE, 90, 91
  WHIFFLE, BABY, 91
Automatic Machine Company
  AUTOMATIC DICE, 138, 207
  EMPIRE, 202
Automatic Machine & Tool Company
  GABEL'S AUTOMATIC ENTERTAINER, 73, 168
  GABEL'S LEADER, 18
  GABEL'S NIAGARA, 19
  GABEL'S TWO-BITS DEWEY, 18
  MERCHANT, THE, 168
Automatic Manufacturing Company
  AUTOMATIC DICE, 139
Automatic Musical Instruments Company (AMI) (AMI/Rowe)
  MODEL A, 83
  MODEL B, 83
  MODEL C, 83
  MODEL E, 84
  MODEL F, 85
  MODEL H, 86
  SINGING TOWERS, 1939, 78
  SINGING TOWERS, 1940, 78
  STREAMLINER, 76
Automatic Sales Company
  TRUE BLUE, 171
  TRUE BLUE, IMPROVED, 171
Automatic Target Machine Company, The
  AUTOMATIC PISTOL, 119
Autosales Corporation
  AUTOSCALE, 188
Bagatelle games, 90
Baker Novelty Company
  BAKER'S PACERS, 53
  ON DECK, 101
Baldecchi Manufacturing Company
  NEVADA CLUB 4 REEL, 67
Bally Manufacturing Company (Corporation), 7, 8, 14, 15, 91, 137, 182
  ADAMS FAMILY, 109
  ALL STAR DELUXE BOWLER, 131
  BABY RESERVE, 161
  BALLY BELL, 51
  BALLY BOOSTER, 101
  BALLYHOO, 91, 92, 93
  BUMPER, 92
  BUS STOP, 199
  CAPT. FANTASTIC, 92
  CLOVER BELL, 59
  DOLLY PATRON, 92, 108
  DRAW BELL, 58
  DRAW BELL, DELUXE, 59
  DRAW BELL, TRIPLE, 60
  FIREBALL, 4
  HIGH HAND, 54
  JUMBO, 98
  LITE-A-PAX, 162
  MONEY HONEY, 9, 65, 208
  PLAYBOY, 108
  RELIANCE, 50, 51
  ROCKET, 92
  SIX MILLION DOLLAR MAN, 108
  STAR TREK, 108
  SUPER BOWLER, 131
  TRIPLE BELL, 58
Banner Specialty Company
  LEADER, 150
Barnes, Monroe, Manufacturer
  CRESCENT CIGAR WHEEL, 141
Barnes Scale Company
  NAVCO, 191
Behn, Henry (inventor), 138
Belk, Schafer & Company
  HUMMER, 165
  PATENT CIGAR SELLER, 165
Bennett, Paul, & Company
  DEUCES WILD, 163
  DOUGHBOY, 163
  LUCKY PACK, 163
  TEST YOUR SKILL, 163
  TOKETTE, 163
Benton Harbor Novelty Company
  IMP, 36, 152
  NEW IMP, 36
Berger, Paul E., Manufacturing Company, 10, 14
  CHICAGO RIDGE, 115
  MAGIC FORTUNE TELLER, THE, 115
  PEANUT VENDER, 165
  SAPHO, 115
Bevco Company, Inc., 167
Bishop, Charles C., & Company
  TRIOGRAPH, THE, 137
  BOUNCERINO (Unknown), 164
Brainerd Baxter Corporation
  VEND-N-SLOT, 66
Brunhoff Manufacturing Company
  DIAMOND EYE, 143
  SPINNING TOP, 144
Bryant Pattern & Novelty Company
  ZODIAC, 145, 146
Buckley Manufacturing Company, 7, 8
  BONES, 50
  CHERRY BONES, 50
  FAVORITE, 92
  KENTUCKY DERBY, 62
  POINTMAKER, ELECTRONIC, CRISS CROSS, 65
  TRACK ODDS, 62
Bull Durham Novelties
  TRIPLE JACK GOLD AWARD, 44
Burtmeier Manufacturing Corporation
  PONY, 36
Caille, Adolph, 12
Caille, Arthur (Art), 12
Caille Brothers Company, The, 8, 18, 69
  ARISTOCRAT, 189
  ARISTOCRAT DELUXE, 189, 190
  AUTOMATIC BEAM SCALE, 186
  BALL GUM VENDER, 152
  BIG SIX, 10
  BON TON, 148, 149
  BULLFROG, 20
  CAIL-O-PHONE, 73
  CAIL-O-SCOPE, 118
  CENTAUR, 20
  CHECK BOY, 24
  COMBINATION BLOW & GRIP, COUNTER, 120
  COMBINATION BLOW & GRIP, ELECTRIC, 120
  DETROIT, NEW CENTURY, 18
  DICTATOR, 46
  DOUGH BOY, 46
  ECLIPSE, 19
  ELK, 155
  FORTUNE VENDER, 136, 153
  HY-LO, 146
  JACK POT kit, 31
  JUNIOR BELL, 136
  JUNIOR BELL STYLE 2, 152
  LIBERTY BELL, 22
  LIBERTY BELL GUM VENDER, 22
  LITTLE MANHATTAN, 90
  LOG CABIN, 91
  MASCOT ELECTRIC TOWER GRIP, 120
  MODERNE, 192
  MYSTIC WHEELS ("Gentlemen"), 121
  O.K. FRONT VENDER, 1923, 26
  O.K. FRONT VENDER, 1924 ("Beachcomber"), 28
  OLYMPIA PUNCHER, 120
  PUCK, 18
  PUCK, NEW CENTURY, 18
  REGISTER, 148
  ROULETTE, 20
  SILVER CUP, 24
  SILVER CUP, ORIGINAL, 24
  SUPERIOR GRAND PRIZE, 37
  SUPERIOR JACKPOT, 1928, 31
  SUPERIOR JACKPOT, 1929, 30
  TIGER, 155
  TRIPLE, 19
  TWIN, BIG SIX-ECLIPSE
  TWO-BITS, 18
  WASHINGTON, 187
  WASP, 145, 146
  WHEEL OF FATE ("Gentlemen"), 120
Caille-Schiemer Company, The, 10
  LITTLE MANHATTAN, 90
  PUCK, 10
Canda, Leo, Company, 140
  GIANT ARROW, 140
  GIANT CARD, 140
  GIANT DICE, 140
  GIANT POLICY, 140
  IMPROVED ROULETTE, 140
  PERFECTION CARD, 143
Capehart Corporation
  MODEL NO.3 JUNIOR, 73
Cavalier Corporation
  C-51 (Coca-Cola®), 182
Cebco Products Company
  DUPLEX, 180
  TRIPLEX, 180
C. & F. Manufacturing Company
  BABY GRAND, 38
Chester-Pollard Amusement Company, Inc.
  PLAY DERBY, 125
  PLAY FOOTBALL, 123, 134
Chicago, Illinois, 7
  "Coin Machine Capital Of The World", 7, 19
Chicago Coin Machine Company, 7, 8
  ALL STAR BASEBALL, 199
  BAND BOX, 84
  BASKETBALL CHAMPION, 199
  BEAM LITE, 97
  DRIVE MASTER, 133
  HIT PARADE, 85
  LELAND STANDARD, 95
  ON DECK, 101
  SCALE-O-MATIC, 203
  SHARPSHOOTER, 134
  SOUND STAGE, 109
  STEAM SHOVEL, 132
  TV FLIPPER, 134
  TV HOCKEY, 134
  TV PING PONG, 134
  TV HOCKEY, 134
Chicago Nickel Works, 138
Chicago Recording Scale Company
  AUTOMATIC RECORDING SCALE, 185
Cinematone Corporation
  PENNY PHONO, 76
  PENNY PHONO CONSOLE, 76
Clark, Norm (pinball designer), 107
Clawson, Clement C. (inventor), 12
Clawson Slot Machine Company, 12
  AUTOMATIC DICE, 138, 153
  IMPROVED ROULETTE, 140
  THREE JACKPOT, 12, 13, 17, 34
Clovena Machinery Company
  BREATH MINTS, 172
Coan Manufacturing Company
  U-SELECT-IT MODEL 74-AP, 181
Coan-Slettland Manufacturing Company, 181
Coca-Cola®, 167, 179, 182, back cover
Coin machines
  coin controlled, 6
  coinage, 13, 14, 15
  Collecting, 9, 69, 70, 91, 92, 112, 165, 183, 186, 197-214
  locations, 166
  price guide, 215-216
  production, 7, 166
  replicas, 12, 204
  restoration, 12, 210-214
  wartime revamps, 92, 127
Colby, John (inventor), 113
Colby Specialty Supply Company
  COMBINATION LUNG TESTER, 113
Collins, Merriam, & Company
  PEERLESS ADVERTISER, 142
Colonial Scale Company
  FAIR WEIGH GOLF, 192
Columbia Phonograph Company
  TYPE A, 70
  TYPE AS GRAPHOPHONE, 70
  TYPE BS EAGLE, 71
Columbia Scale Company
  HONEST WEIGHT, 190
Columbia Weighing Machine Company
  COLUMBIA MIRROR, 190
  COLUMBIA MIRROR CHIMES, 190
  HONEST WEIGHT, 190
Columbian Novelty Company
  SLOT MACHINE, 12
Columbus Vending Machine Company
  MODEL 34 GAMBLER, 179
Consolidated Automatic Merchandising Corporation (C.A.M.Co.), 192, 194
  LITTLE GIANT, 192
Continental Novelty Company
  PERFUME, 172
Cordray, Henry (inventor), 138
Cowper Manufacturing Company, 10
  DRAW POKER, 148
  OREGON, 14
Cox, Palmer ("Brownies"), 17
Crystal Vending Company
  CRYSTALETS, 173
Data East Pinball (Sega)
  LETHAL WEAPON III, 108
  MICHAEL JORDAN, 108
Daval Manufacturing Company
  ACE, 162
  AMERICAN EAGLE, 162, 163
  AMERICAN EAGLE, STAR, 162
  BEST HAND, 163
  CUB, 162
  DAVAL GUM VENDER, 156
  DAVAL GUM VENDER JACKPOT, 156
  DAVAL GUM VENDER JACKPOT, ULTRA MODERN, 156
  MARVEL, 162, 163
  MEXICAN BASEBALL, 163
  OOMPH, 163
  SMOKE REELS, 137
Decatur Fairest Wheel Company
  FAIREST WHEEL, 142
  FAIREST WHEEL NO. 3, 143
Dennison, John, Mechanical Model Maker
  MUSICAL FAIRY, 110
  SORCERER FORTUNE TELLER, 111
Dietz, Andrew T.
  SELF SERVICE, 176
Dillon Manufacturing Company
  AMERICAN POSTMASTER, 181
Donne, Leon (inventor), 112
Doraldina Corporation
  PRINCESS DORALDINA, 124
Dr. Pepper®, 167, 182
Drobisch Brothers & Company
  ADVERTISING REGISTER, 142
  LEADER, THE, 142
  STAR ADVERTISER, FRONT COVER, 2
  VICTOR, 141
Duke, James, 167
Dual Parking Meter Company, 181
DuGrenier, Inc., 181
DuGrenier, Arthur H., Inc.
  ADAMS GUM, 181
Duncan Industries
  DUNCAN METER MODEL 80, 182
Durant, Lyn (designer), 103, 129
Dynamo Corporation
  PRO-BUILT SOCCER, 134
Eagle Manufacturing Company
  EAGLE, 22
  EAGLE GUM VENDER, 22
E.A.M. (unknown dicer), 203
Edison Phonograph Works
  ACME, 70
  AUTOMATIC PHONOGRAPH, 69
  MODEL M, 69
Edwards Novelty Company
  GRIP TESTER, 124
Engel Manufacturing Company
  IMP, 152
Ennis & Carr
  PERFECTION, 147
Evans, H.C., & Company, 8
Everitt, Percival (inventor), 6, 183, 184
  LIEBES WAAGE, 184
Excelsior Race Track Company
  EXCELSIOR, 138, 204
Exhibit Supply Company, 7, 8, 163
  ARCADE DICE FORTUNE TELLER, 153
  BIG BRONCHO, 131
  CHUCK-A-LETTE, 48
  CRYSTAL PALACE, 127
  ELECTRIC EYE, 126, 127
  ELECTRIC EYE, JACKPOT, 127
  IRON CLAW, 126
  JOCKEY CLUB, 48
  MONARCH MERCHANDISER, 128
  ORACLE, 122
  PLAY BALL, 93
  RED DOG, 160
  SMOKY, 102
  TALLY HO, 103
  20TH CENTURY, 103
  VITALIZER, 203
Fairbanks, E. & T., Company, 6, 183
  WEIGHING MACHINE, 183
Fantasy machines, 12, 204
Fanzines (collector magazines), 9, 209
Farrington Cigar Company
  GRIP TEST, 112
Fey, Charles August, 10, 151, 153, 160, 188
Fey, Chas., & Company
  BIG SIX GUM VENDER, 24
  LIBERTY BELL, 4, 10, 11
  ON THE LEVEL, 149
  SKILL ROLL, 150, 151
  TRIPLE ROLL, 150, 151
  TWIN JAK-POT, 44
Fey, Edmund C.
  ACE, THE, 153
Field Paper Products (Manufacturing) Company, 38
  DING THE DINGER, 154
  4 JACKS, 34
  INDOOR SPORTS, 154
  POKER-E-NO, 154
Fleer, Frank H., & Company
  TRY A SAMPLE, 167
Flour City Manufacturing Company, 138
Fort Wayne Novelty Manufacturing Company
  LITTLE JOE, 158
Freeport Novelty Company
  GOO GOO, 168
  SODA MINT, 168
Fuller, Paul (designer), 78, 80, 81, 82
Gabel, John (inventor), 18
Gabel, The John, Manufacturing Company, 69
  AUTOMATIC ENTERTAINER, 1929, 72
  GABEL'S AUTOMATIC ENTERTAINER, 1917, 73
  JUNIOR ELITE, 76
Game Plan, Inc., 109
Games, Inc.,
  HUNTER, 64
Gatter, Rudolph (inventor), 122
Gatter Novelty Company
  AUTOMATIC BOWLING ALLEY, 122
  TEN PIN GUM VENDER, 122
Genco, Inc.
  BASE BALL, 97
  BOWLING LEAGUE, 130
  CHEER LEADER, 100
  400, 104
  SPIT FIRE, 99
  TRIPLE ACTION, 103
  WHIZZ, 104
General Coin Automatic Company
  TYPE-O-METER, 178
Golden Nugget Casino, 12, 13
  GOLDEN NUGGET, 12, 204
Gottlieb, D., & Company, 7, 8, 91
  ARABIAN KNIGHTS, 104
  ELECTRIC SCOREBOARD, 101
  FIVE STAR FINAL, 93
  FLIPPER, 106
  HI-DIVER, 105
  HUMPTY DUMPTY, 92, 103
  KEWPIE DOLL, 106
  NEW DAILY RACES, 102
  PIN WHEEL, 105
  QUEEN OF DIAMONDS, 106
  RACK-A-BALL, 199
  SUNSHINE DERBY, 99
Gravity Vending Machine Company
  GUM MACHINE, 171
  HONEY BREATH BALLS, 165, 171
Great States Manufacturing Company
  BUCK-A-DAY, 156
  FOOT BALL PRACTICE, 155, 156
  HAPPY DAYS, 156
  TARGET PRACTICE, 155
Griffin Vending Company
  GRIFFIN, 175
Griswold, M. O., & Company
  PENNY PEANUT, 174
  RED STAR, 175
  STAR, 148
Groetchen Tool (& Manufacturing) Company, 8, 137
  ATOM, 162
  BARTENDER, THE, 159
  COLUMBIA, 49
  COLUMBIA, GOLD AWARD, 53
  COLUMBIA STANDARD, 52
  HIGH TENSION, 160
  IMP, 162
  KLIX, 163
  KILL THE JAP, 127
  LIBERTY, 163
  POK-O-REEL, 154
  POK-O-REEL GUM VENDER, 154
  POK-O-REEL, 1941, 163
  TAVERN STYLE A, 159
  TAVERN STYLE B, 159, 160
  21 VENDER, 157
  WINGS, 163
  YANKEE, 163
Gross, Rube (designer), 92
Gross, Rube, & Company
  FURY, 98
  TORPEDO, 98
Hamilton Manufacturing Company
  DAISY, 148
Hance Manufacturing Company, 176
  AUTO, 177
  O.I.C. VENDER, 202
  REX, 176
  TARGET PRACTICE, 121
  TARGET PRACTICE, PENNY BACK, 121
Hanson Scale Company
  PENNY SCALE, 195
Harris, Anthony (inventor), 6
Harvard Automatic Machine Company
  METAL STAMPER, 122
Health Lift & Exercising Machine Company
  HEALTH LIFT, 112
Heene Automatic Stamping Machine Company, 122
Hilo Gum Company
  HILO GUM CLIMAX, 171
Hochriem, Gustave (inventor), 10, 13
Hoff Vending Corporation
  WRIGLEY'S NO.5A, 178
Holcomb & Hoke Manufacturing Company
  ELECTRAMUSE GRAND, 73
  ELECTRAMUSE SUPER-TONE, 73
Hollands, Jet
  STAR PEPSIN, 167
Hopkins & Kost Company, Ltd.
  FLEUR DE LIS, 173
Houston Showcase & Manufacturing Company
  SWEET HEART, 94
  SWEET HEART DOUBLE, 94
Hutchison Engineering Company
  SCRAM, 93
Ice machine, 203
Ideal Weighing Machine Company
  IDEAL SCALE, 191
Illinois Machine Company, 10
  PUCK, 10
In And Out-Door Games Company
  WHOOPEE, 90
Industrial Carton Coolers
  CARTON COOLER, 179
Industry Novelty Company, 7
International Devices Corporation
  ARISTOCRAT, 193
International Game Technology (IGT), 14, 16
International Mutoscope Reel Company
  ALL STEEL MUTOSCOPE, 123
  CHAMPION, 126
  DRIVEMOBILE, 133
  DROP KICK, 128
  PHOTOMATIC, 130
International Ticket Scale Corporation
  INTERNATIONAL TICKET SCALE, MODEL A, 191
International Vending Company
  SAFETY MATCHES, 171
Inwersen Manufacturing Company, 90
  HOME PARLOR KEEPS, 90, 125
Iowa Novelty Company
  O.K. MINT VENDER, 25
Iowa Paper Company
  MATCHES, 174
Jackson Supply Company
  HONEST CLERK, THE, 172
Jackson Vending Company
  SAFETY MATCHES, 171
Jacobs Novelty Company
  616 ILLUMINATED FRONT, 77
Jennings, O.D., & Company
  AUTOMATIC COUNTER MINT VENDER, 26
  AUTOMATIC MINT VENDER, 26
  BASEBALL, 32
  BOBTAIL, 55
  BRONZE CHIEF, 55
  BUCKAROO, 64
  CARD MACHINE, 157
  CENTURY CLUB TRIPL-JACK, 45
  CENTURY TRIPL-JACK, IMPROVED, 44
  CHALLENGER, 59
  CHALLENGER, MONTE CARLO, 59
  CHIEF, 50
  CHIEF VENDER SKILL, 53
  DUCHESS DOUBL-JACK VENDER, 44
  ELDORADO, 66
  EXPORT CHIEF, 63
  FAST TIME, 54
  FAST TIME FREE PLAY, 54
  GALAXY, 67
  GALAXY MECHANICAL JACKPOT, 67
  GOLF BALL VENDER, 47
  GOOD LUCK, 54
  GOOD LUCK DELUXE, 54
  GOVERNOR, 66
  GRUB STAKE, 66
  HARVEST MOON, 54
  JOKER, BRONZE CHIEF, 64
  LITTLE DUKE, 41, 544
  LITTLE DUKE JACKPOT, SELECTIVE, 41
  LITTLE DUKE TRIPL-JACK VENDER, SELECTIVE, 42
  LITTLE MERCHANT, 157
  LUCKY PUNTER, 66
  OPERATORS BELL, 25
  PENNY CLUB, 161
  PERFECTED JACKPOT, 32
  POKER GAME, 157
  PRIMER JACKPOT, 31
  RAINBOW, 37
  REBATER, 157
  ROCKAWAY, 34, 37

Page	Pos	Item	Price
162	BL	Daval AMERICAN EAGLE	200
162	TR	Groetchen IMP	150
162	CR	Daval CUB	130
163	BR	Bennett TEST YOUR SKILL	250
163	CL	Groetchen YANKEE	200
163	TR	Daval BEST HAND	350
163	BR	Arkansas 3-A-LIKE	250
164	TL	Atlas TILT TEST	170
164	BL	Shelby LUCKY 7	220
164	TC	Maitland MYSTIC DERBY	3500
164	CR	BOUNCERINO	220
165	CL	Belk, Schafer HUMMER	3700
166	C	Volkman, Stollwerck CONFECTIONS	3200
166	CR	U. S. Slot ADAMS PEPSIN	2800
167	CL	Fleer TRY A SAMPLE	2600
167	BR	Hollands STAR PEPSIN	2000
168	BL	Automatic Machine MERCHANT	25000
168	C	Regina GUM VENDER	7500
168	TR	Freeport SODA MINT GUM	1500
168	BR	National COLGAN'S GUM	1800
169	TL	Automatic AUTOMATIC CLERK	600
169	CL	Winsted POTTER'S BEST	2500
169	CR	Pulver TOO CHOOS	1200
170	TL	Price COLLAR BUTTON	1500
170	BL	Austin, Nichols FOXY GRANDPA	2800
170	CR	National TAFFY TOLU	1450
171	TL	National Automatic IMPROVED TRUE BLUE	2200
171	BL	Gravity GUM	2200
171	TC	International SAFETY MATCHES	1200
171	TR	Advance CLIMAX CHICAGO	600
171	BR	HILO GUM CLIMAX	900
172	TL	Continental PERFUME	2600
172	BL	Pope AUTOMATIC CIGAR	3000
172	TR	Jackson HONEST CLERK	3200
172	BR	Clovena BREATH MINTS	3800
173	TL	Crystal CRYSTALETS	1250
173	BL	Hopkins & Kost FLEUR DE LIS	3200
173	TR	Regina STYLE 18 GUM	7500
174	TL	Griswold PENNY PEANUT	900
174	BL	Advance GLOBE	600
174	CR	Iowa Paper MATCHES	800
175	TL	National BREATH PERFUME	1250
175	BL	Winters RED STAR	4000
175	TC	Griffin Vending GRIFFIN	30000
175	CR	Northwestern SELLUM	350
176	TL	LINCOLN HOT NUT	800
176	BL	Dietz SELF SERVICE	850
176	TC	Hance REX	1000
176	TR	National THE NATIONAL	250
176	BR	Weeks PENCIL VENDER	500
177	CL	Krema MATCH VENDER	350
177	TR	Hance AUTO	700
177	BR	Stanley HOT NUT	1750
178	TL	Ad-Lee E-Z BASEBALL	700
178	BL	Rowe DELUXE	450
178	TC	Hoff WRIGLEY'S 5A	350
178	TR	D. Robbins EMPIRE	450
178	BR	General Coin TYPE-O-METER	1200
179	TL	Columbus 34 GAMBLER	300
179	BL	Industrial CARTON COOLER	1200
179	TR	Mills Automatic MERCHANT	250
179	BR	Northwestern 33 PENNY BACK	550
180	TL	Los Angeles SUN	30
180	BL	Lawrence ROL-A-COIN	120
180	TC	Silver King SILVER KING	100
180	TR	Stoner NEW UNIVENDER	150
180	BR	Tropical Trading CHALLENGER	250
181	TL	Magee-Hale PARK-O-METER	30
181	BL	Dillon AMERICAN POSTMASTER	100
181	TC	DuGrenier ADAMS GUM	60
181	TR	Coan U-SELECT-IT	180
182	TL	Stoner CANDY UNIVENDER	180
182	BL	Vendo V-81	1500
182	C	Duncan DUNCAN METER	25
182	CR	Cavalier C-51	600
183	CR	E. & T. Fairbanks WEIGHING MACHINE	15000
184	R	Everitt LIEBES-WAAGE	15000
185	TL	Chicago Recording RECORDING SCALE	16000
185	CL	National Automatic NATIONAL	1800
185	TC	National Weighing NATIONAL	1800
185	TR	Watling STYLE 1 GUESSING	4500
185	BR	National Novelty NATIONAL	2000
186	CR	Watling STYLE 4 WEIGHING	2500
186	BR	Caille AUTOMATIC BEAM	2400
187	TL	Mills WEIGHING MACHINE	1600
187	TC	Caille WASHINGTON	8500
187	TR	United Vending WATERPROOF TALKING	9000
187	BL	Mutual HONEST WEIGHT	2250
188	TL	Pacific PUBLIC WEIGHER	2000
188	BL	Automatic Beam AUTO	2200
188	TC	Watling Scale MODEL 5 STEEL	1200
188	C	Autosales AUTOSCALE	2000
188	CR	Mills 1918 ACCURATE	1600
189	TL	Peerless ARISTOCRAT	1250
189	BL	National Novelty MODEL 103	1600
189	TC	Salter NO. 500 catalog page	200
189	CR	Peerless ARISTOCRAT DELUXE	2500
190	TL	Columbia MIRROR	1200
190	TC	Columbia HONEST WEIGHT	1250
190	TR	Peerless ARISTOCRAT JUNIOR	900
190	B	Watling GAMBLER	2200
191	TL	National Automatic NORMANY CHIMES	1800
191	TC	International MODEL A TICKET	750
191	TR	Public WEIGHT VERIFIER	3200
191	BL	Ideal IDEAL	225
191	BR	Barnes NAVCO	300
192	TL	Caille MODERNE	300
192	TC	Colonial FAIRWAY GOLF	6500
192	TR	Camco LITTLE GIANT	400
192	BL	Mills MODERN	300
192	BR	Mills MODERN WITH CHART model	250
193	TL	Rock-Ola LOBOY MODEL B	250
193	TC	International Devices ARISTOCRAT	250
193	TR	Watling TOM THUMB	300
193	BL	A.S.L. Sales ACCURATE	120
193	BR	American 300	250
194	TL	Raizner ADVERTISING SCALE	650
194	TR	Peerless BANTAM	250
194	BL	Kirk GUESS-ER	800
194	BC	Public GUESS	600
194	BR	Public PERSONAL	150
195	TLC	American 403	120
195	TRC	Watling FORTUNE	450
195	BL	Hanson PENNY	150
195	BC	American 400	100
195	BR	Watling BIG GUESSING	1600
196	TL	American 530	100
196	C	Watling SUPER HOROSCOPE	600
196	TR	Weight Check HONEST WEIGHER	150
202	CL	Hance O.I.C. VENDER	12500
202	CR	Automatic Machine EMPIRE	4500
202	BR	Unknown	2800
203	TL	Exhibit VITALIZER	250
203	TC	E.A.M. 9500	
203	TR	Chicago Coin SCALE-O-MATIC	450
203	BL	AIR	80
203	BC	ICE	No value
203	BR	NEVADA toy bank	40
204	TL	Reproduction GOLDEN NUGGET	1600
204	CR	Reproduction 3 JACKS	350
206	BR	Pace DANDY VENDER	350
207	T]	Jennings SUN CHIEF	1600
208	T	Bally 742A MONEY HONEY	1800
Back cover TR		U. S. Novelty WINNER	1400
Back cover C		Millard GOLD SEAL	150
Back cover BC		Vendo V-56	1200
Back cover BR		Pulver CHOCOLATE COCOA	4200

Other Titles from Schiffer Publishing

Jukeboxes Michael Adams, Jürgen Lukas, and Thomas Maschke. Here is the history of the development, the technology, and the stars of jukeboxes. Here also is information on the manufacturers, a photo chronicle of their evolution, and advice on collecting, maintaining and restoring jukeboxes from the '50s and '60s plus an illustrated index of model numbers by manufacturer.
Size: 8 1/2" x 11"   305 photos   144pp.
ISBN: 0-88740-876-1   hard cover $35.00

Guide to Vintage Trade Stimulators & Counter Games Dick Beuschel. You'll get to know all the makers and their machines by name, date, and appearance in this colorful collector's guide. Photos, detailed machine descriptions, manufacturer production data, an interchangable machine name list and price guide.
Size: 8 1/2" x 11"   816 photos   196 pp.
ISBN: 0-7643-0119-5   hard cover $34.95

Arcade Treasures Bill Kurtz. This book is filled with color photos of the greatest games ever to hit the arcade scene—the earliest strength testers and fortune tellers, wild pinball games from mid-century, and modern electronic video games. Also discusses novelty products, game trends, and a history of significant manufacturers and designers.
Size:8 1/2" x 11"   447 photos   176 pp.
Price Guide
ISBN: 0-88740-619-X   hard cover $39.95

Pinball Machines (Second Edition) Heribert Eiden & Jürgen Lukas. A collector's book as well as a pinball gallery of the op art genre, this is a chronology, from the simple early mechanical pinball machines up to their modern fully-electronic equivalents. It provides an overview of all the variety, the artistry, and the manufacturing ingenuity that have gone into pinball machines over the years.
Size: 8 1/2" x 11"   Price Guide   168 pp.
ISBN: 0-7643-0316-3   hard cover $35.00

The Musical Box Dr. Arthur W.J.G. Ord-Hume. Contains diagrams of the various types of mechanical movements used over the centuries, indexes of manufacturers, box styles, and tune sheets, complete American and British patent lists, and tips on maintaining, repairing, and purchasing boxes. Many of the loveliest musical boxes ever crafted are illustrated in full color in this book, with hundreds more black and white photos showing their mechanical workings. Eighty detailed diagrams explaining their functions help collectors use, maintain and repair the boxes in their collections.
Size: 8 1/2" x 11"   600 illustrations   336 pp.
Price Guide
ISBN: 0-88740-764-1   hard cover $79.95

Philco Radio: 1928-1942 Ron Ramirez, with Michael Prosise A superb reference book on Philco, the leading radio manufacturer during radio's "Golden Age." Specifications for each model given. A year-by-year look at Philco's radio line each year between 1928, when the company began to make radios, and 1942, when World War II put a halt to radio production.
Size: 8 1/2" x 11"   828 photos & drawings 160 pp.
Price Guide
ISBN: 0-88740-547-9   soft cover $29.95

Radios by Hallicrafters Chuck Dachis. This book includes over 1000 photographs of radio receivers, transmitters, and speakers, early television sets, electronics accessories and advertising material produced by this Chicago-based firm. Technical descriptions are provided for every known Hallicrafters model, including dates of production, model numbers, accompanying pieces, and original prices.
Size: 8 1/2" x 11"   1000 photos   225 pp.
Price Guide
ISBN: 0-88740-929-6   soft cover $29.95

Zenith Radio: The Early Years 1919-1935 Harold Cones, Ph.D, and John Bryant, AIA, with Martin Blankinship and William Wade. Presents the documented story of Zenith radio and company from 1919 through 1935. Tells the story of Zenith's impact on early radio history with photographs, documents, and information, as well as color portraits of many Zenith radios of the era.
Size: 8 1/2" x 11"   Value Guide   192 pp.
ISBN: 0-7643-0367-8   soft cover $29.95

Zenith Transistor Radios Evolution of a Classic Norman R. Smith. This comprehensive book includes a complete listing of all transistor radio models created by Zenith from 1955 through 1965. Outstanding color photos from original Zenith sales sheets as well as information on each model are presented in a year by year order of production. Never before published photographs, documents, and original drawings from the Zenith archives, as well as a large collection of original Zenith advertising, fill the pages.
Size: 8 1/2" x 11"   252 photos   160 pp.
Price Guide
ISBN: 0-7643-0015-6   soft cover $29.95

The Zenith TRANS-OCEANIC, The Royalty of Radios. John H. Bryant, AIA and Harold N. Cones, Ph.D. With access to the Zenith corporate archives, this book presents the engrossing stories of the development and use of the Trans-Oceanic throughout its forty year life. They present a wealth of photos, documents and information concerning these fascinating radios, their collection, preservation and restoration.
Size: 8 1/2" x 11"   180 photos   128pp.
Value Guide
ISBN: 0-88740-708-0   soft cover $24.95

Genuine Plastic Radios of the Mid-Century Ken Jupp and Leslie Piña. Mid-twentieth century table radios made primarily of brightly colored plastic represent a relative newcomer to the radio collecting arena. These examples of American industrial design and popular culture can be found at flea markets, garage and house sales. With color photographs of radios, plus advertisements and black and white vintage photos, this pioneering book is a must for anyone interested in radios, mid-century industrial design, or popular culture.
Size: 11" x 8 1/2"   439 color photos   208 pp.
Price Guide   Index
ISBN: 0-7643-0108-X   hard cover $39.95

The Talking Machine: An Illustrated Compendium 1877-1919 Timothy C. Fabrizio & George F. Paul. This thorough compendium contains color photos showing an incredible variety of external-horn as well as internal-horn talking machines. The authoritative text and up-to-date value guide complement a wealth of visuals, providing a veritable library.
8 1/2" x 11"   553 photos   256 pp.
ISBN: 0-7643-0241-8   hard cover $69.95

Telephone Collecting: Seven Decades of Design Kate E. Dooner. Here, in text and over 250 color photographs, a history of the design of the telephone is presented from Art Deco years to novelty phones of the 1980s. The largest telephone companies are discussed, including Western Electric, Automatic Electric, Stromberg-Carlson, Kellogg, and North Electric. Plus there's information on European telephones.
Size: 8 1/2" x 11"   Price Guide   128 pp.
ISBN: 0-88740-489-8   soft cover $24.95

Collectible Novelty Phones If Mr. Bell Could See Me Now James David Davis. Here is a guide to more than two hundred novelty telephones, as well as miniature and toy telephones. Character telephones, telephone advertising products, and figural-type phones are all in demand. Starting in the '70s, these telephones were produced to promote company products, television shows, comic strips, movies, and sports. Miniature and toy telephones are highly desired by many antique telephone collectors and both are included in this enjoyable book. Descriptive captions list each phone's specific features.
Size: 8 1/2" x 11"   315 color photos   144 pp.
Price Guide
ISBN: 0-7643-0472-0   soft cover $24.95

One Hundred Years of Bell Telephone Richard D. Mountjoy. From the Coffin sets of the 1870s to the Princess phones of the 1960s and beyond, this book explores the technology and the history of the telephone. This definitive work provides detailed information which will help identify a piece and will take the guess work out of dating equipment. For those who are restoring a telephone and would like to ensure its historical accuracy, this book will make it easy to match pieces correctly.
Size: 8 1/2" x 11"   350 color photos   176pp.
Price Guide
ISBN: 0-88740-872-9   soft cover $29.95

Telephones: Antique to Modern *2nd Edition* Kate Dooner. Enjoy this wonderful depiction of innovative telephones in over 500 color photographs tracing the development of the telephone from Bell's first experimental equipment. The book pictures exquisite examples of wooden box phones, vanities, upright "candlesticks," and desk stand or "cradle" phones, including some Canadian and European models. This volume has become an important reference for the novice and avid telephone collector alike, with full descriptions of the numerous telephone companies and manufacturers and, in this 2nd edition, an updated value guide. This is the most expansive work ever compiled on collecting antique telephones, helping it become a growing and exciting hobby.
Size: 8 1/2" x 11"   500 Photos   176 pp.
Value Guide
ISBN: 0-7643-0352-X   soft cover $29.95

Metal Toys from Nuremberg, 1910-1979 Gerhard Walter. The book covers an important era in toy manufacturing with wonderfully detailed color and black-and-white photographs. Provides an exact record of manufacturing dates, catalog specifications, patent and registration papers, market values, and accurate details of construction materials.
Size: 7 3/4" x 10 1/2"   Price Guide   144 pp.
ISBN: 0-88740-435-9   hard cover $35.00

Battery Toys Brian Moran. A comprehensive view of the world of battery toys. Battery-operated robots, animals, people, character toys, banks, monsters, and cars are shown, compared, and analyzed. Vehicles and figurals from the mid-1950s and the 1960s are the major portion of this book, with some choice examples from the late 1940s and the 1970s. Detailed desirability chart.
Size: 8 1/2" x 11"   153 color photos   192 pp.
ISBN: 0-88740-003-5   hard cover $29.50

Steam Toys: A Symphony In Motion Morton Hirschberg. Hooked to a steam engine or operated by hand, these pieces *move*—mills grind, blacksmiths hammer, Ferris wheels and carousels revolve, woodsmen chop, and mistrels dance. Almost 600 full-color photographs document the work of 25 major manufacturers, with illustrations of the marks they used. Dimensions and years of production are discussed for each piece.
Size: 8 1/2" x 11"   600 photos   256 pp.
Price Guide   Index
ISBN: 0-7643-0009-1   hard cover $69.95

Puppets & Marionettes: A Collec-tor's Handbook & Price Guide Jan Lindenberger. From Pinocchio to Kermit the Frog, Indian stick figures to Jerry Mahoney puppets, these collectibles have entertained and educated generations. *Puppets & Marionettes* will bring back memories of childhood, of the early days of television and the hours of delight these animated characters have brought into our lives.
Size: 6" x 9"   560 color prints   176 pp.
Price Guide
ISBN: 0-7643-0279-5   soft cover $19.95

A Penny Saved: Still and Mechanical Banks Don Duer. The history of penny banks is illustrated with these banks made of cast iron, tin, pottery, wood, and pot metal in forms that include folk art, political events, and special places. Color photos present intriguing American banks arranged from the 18th century through the present.
Size: 8 1/2" x 11"   292 color photos   176 pp.
Price Guide
ISBN: 0-88740-528-2   hard cover $59.95

The Penny Bank Book, Revised Andy and Susan Moore. 1,670 still banks are beautifully presented in color and black-and-white illustrations. Focuses on American and English banks, with examples of still banks from other countries. Includes histories, details about the banks, and information about clubs and publications.
Size: 9" x 12"   Price Guide   192 pp.
ISBN: 0-7643-0377-5   hard cover $49.95

Baby Boomer Toys and Collectibles Carol Turpen. Color photos of over 780 toys and a price guide make this a necessary reference for collectors and dealers. Covers favorites from robots, space toys to race cars, Hanna-Barbera toys, and Beatles collectibles.
Size: 8 1/2" x 11"   495 color photos   204 pp.
Price Guide
ISBN: 0-88740-495-2   soft cover $29.95

About the Author

Richard Bueschel states that the purpose of his book is to "bring some of the charm and excitement of coin machine history, identification and collecting to a broader scope of enthusiasts..."; a task he has easily and enjoyably accomplished. Mr. Bueschel has one of the largest collections of historical materials related to coin operated machines. He is an historian, a columnist for *Antique Week* and the editor of the "fanzine" *Coin Op-Classics*. He lives with his wife in Mt. Prospect, Illinois and has two daughters who are avid antiquers in their own right.

BACK COVER PHOTOS

Top Right: United States Novelty WINNER, 1893. The Chicago contribution to dicing for cigars was a cast iron cabinet assembled from side and top flat castings. United States Novelty Company was located on the far south side of Chicago, an unusual location for a chance machine maker in the 1890s as it was in a residential area away from the industrial centers. This machine had a long life. In 1907 it was acquired by The Callie Bros. Co. and produced for years as WINNER DICE. After 1920 it was picked up by makers in three states and continued in aluminum. $1,410.

Center Left: Millard GOLD SEAL CONFECTIONS, 1922. Breath pellets, or multi-colored mint "pills," were very popular vendables, with the machines placed in drug stores and in the bucket shops that sold illegal beer or whiskey. Millard's Gum Vending Corporation of New York City converted their patented gumball vender into a bulk mint vender and in the process dropped the word "Gum" from their corporate name. A larger GOLD SEAL CHEWING GUM version was also made. $160.

Bottom Left: Vendo COCA COLA MODEL V-56, 1958. Essentially a smaller Vendo COCA COLA MODEL V-81, the V-56 was for lower density locations. Carries 56 bottles in all three sizes. Many of the machines had a solid door so you had to open it to see the available selection. Collectors like the V-56 for its selectability and size. They also tend to favor overreproductions with added decals and graphics, as shown. While it may add to the appeal, it ain't real. $1,210.

Bottom Right: Pulver CHOCOLATE COCOA, 1899. An early appreciator of the value of automatic merchandising, the Pulver Chemical Company of Rochester, New York, started making wall vending machines with moving dolls inside in 1897. The word "Chemical" was a lousy name for a gum company, so they changed it to Pulver Chocolate & Chicle Manufacturing Company. The combination metal and wood cabinet CHOCOLATE COCOA has Uncle Sam as the inside figure. $4,220.